Church and Worship Music

Routledge Music Bibliographies

Series Editor: Brad Eden

COMPOSERS

Isaac Albéniz (1998)
Walter A. Clark

C. P. E. Bach (2002)
Doris Bosworth Powers

Samuel Barber (2001)
Wayne C. Wentzel

Béla Bartók, Second Edition (1997)
Elliott Antokoletz

Vincenzo Bellini (2002)
Stephen A. Willier

Alban Berg (1996)
Bryan R. Simms

Leonard Bernstein (2001)
Paul F. Laird

Johannes Brahms (2003)
Heather Platt

Benjamin Britten (1996)
Peter J. Hodgson

William Byrd, Second Edition (2005)
Richard Turbet

Elliott Carter (2000)
John L. Link

Carlos Chávez (1998)
Robert Parker

Frédéric Chopin (1999)
William Smialek

Aaron Copland (2001)
Marta Robertson and Robin Armstrong

Frederick Delius (2005)
Mary Christison Huismann

Gaetano Donizetti (2000)
James P. Cassaro

Edward Elgar (1993)
Christopher Kent

Gabriel Fauré (1999)
Edward R. Phillips

Christoph Willibald Gluck, Second Edition (2003)
Patricia Howard

G. F. Handel, Second Edition (2005)
Mary Ann Parker

Paul Hindemith (2005)
Stephen Luttmann

Charles Ives (2002)
Gayle Sherwood

Scott Joplin (1998)
Nancy R. Ping-Robbins

Zoltán Kodály (1998)
Mícheál Houlahan and Philip Tacka

Franz Liszt, Second Edition (2003)
Michael Saffle

Guillaume de Machaut (1995)
Lawrence Earp

Felix Mendelssohn Bartholdy (2001)
John Michael Cooper

Giovanni Pierluigi da Palestrina (2001)
Clara Marvin

Giacomo Puccini (1999)
Linda B. Fairtile

Maurice Ravel (2004)
Stephen Zank

Gioachino Rossini (2002)
Denise P. Gallo

Camille Saint-Saëns (2003)
Timothy S. Flynn

Alessandro and Domenico Scarlatti (1993)
Carole F. Vidali

Heinrich Schenker (2003)
Benjamin Ayotte

Jean Sibelius (1998)
Glenda D. Goss

Giuseppe Verdi (1998)
Gregory Harwood

Tomás Luis de Victoria (1998)
Eugene Casjen Cramer

Richard Wagner (2002)
Michael Saffle

Adrian Willaert (2004)
David M. Kidger

GENRES

American Music Librarianship (2005)
Carol June Bradley

Central European Folk Music (1996)
Philip V. Bohlman

Chamber Music (2002)
John H. Baron

Choral Music (2001)
Avery T. Sharp and James Michael Floyd

Church and Worship Music (2005)
Avery T. Sharp and James Michael Floyd

Ethnomusicology (2003)
Jennifer C. Post

Jazz Research and Pedagogy, Third Edition (2005)
Eddie S. Meadows

Music in Canada (1997)
Carl Morey

The Musical (2004)
William Everett

North American Indian Music (1997)
Richard Keeling

Opera, Second Edition (2001)
Guy Marco

The Recorder, Second Edition (2003)
Richard Griscom and David Lasocki

Serial Music and Serialism (2001)
John D. Vander Weg

String Quartets (2005)
Mara E. Parker

Women in Music (2005)
Karin Pendle

CHURCH AND WORSHIP MUSIC

AN ANNOTATED BIBLIOGRAPHY OF CONTEMPORARY SCHOLARSHIP

A RESEARCH AND INFORMATION GUIDE

AVERY T. SHARP and JAMES MICHAEL FLOYD

ROUTLEDGE MUSIC BIBLIOGRAPHIES

Published in 2005 by
Routledge
Taylor & Francis Group
270 Madison Avenue
New York, NY 10016

Published in Great Britain by
Routledge
Taylor & Francis Group
2 Park Square
Milton Park, Abingdon
Oxon OX14 4RN

Printed in the United States of America on acid-free paper
10 9 8 7 6 5 4 3 2 1

International Standard Book Number-10: 0-415-96647-7 (Hardcover)
International Standard Book Number-13: 978-0-415-96647-4 (Hardcover)

Library of Congress Cataloging-in-Publication Data

Catalog record is available from the Library of Congress

Taylor & Francis Group
is the Academic Division of T&F Informa plc.

Visit the Taylor & Francis Web site at
http://www.taylorandfrancis.com

and the Routledge Web site at
http://www.routledge-ny.com

Acknowledgments

We sincerely appreciate the assistance of the following individuals and institutions: Kenneth L. Carriveau, Jr., Head, Extended Library Services Unit, and the staff of the Interlibrary Services Department, Baylor University; the staff of the Waco-McLennan County Libraries; the interlibrary loan librarians and staffs at no less than one hundred other academic libraries around the United States; Donna Kennedy, ITS–Information Systems & Services Department, Baylor University; Danielle Blackstone, Crouch Fine Arts Library, Baylor University; and Karen R. Little, indexer of Music Library Association *Notes,* for advice on indexing.

Avery T. Sharp
James Michael Floyd

Contents

Preface

Constant experimentation, change, and turmoil characterized church music in the United States during the period from circa 1960 to the end of the twentieth century. At times, bitter "battles" raged between diehard sacred music "traditionalists," contemporary church music "radicals," and all the millions of church leaders, church musicians, and church members whose preferences and opinions fell somewhere between the two extremes. Disputes regarding church music were an important part of the churches' "worship wars" during the period. In the introduction to his book *America's Worship Wars,* the former music minister, pastor, and, currently, seminary professor Terry W. York states:

> The theo-political "wars" experienced inside the confines of some of America's major Christian denominations are but subsets of the nation's larger "culture wars." . . . It seems an awkward juxtaposition to include the words "worship" and "wars" in the same sentence or title. Yet, strangely, the two events have something in common. Both worship and war engage the extreme depth of the human soul, even though they are poles apart in human behavior. . . . In researching the dynamics and discourse of change that took place . . . one often encounters words such as war, battle, kill, camps, tribes, strategies . . . revolution, battlefield, and casualties.

1. SCOPE

The online Hyperdictionary (www.hyperdictionary.com) defines "church music" as "a genre of music composed for performance as part of religious ceremonies." That description is now too limited and traditional to be accurate. Much "traditional sacred music," "religious folk music," and "contemporary Christian music" is performed outside "religious ceremonies." Performance venues of "traditional sacred music" and "religious folk music" are more likely now to be choral concerts in schools, colleges, and universities, in addition to community choruses and professional ensembles. "Contemporary Christian music" is performed as often in concerts as in religious services.

Coverage in this compilation is restricted to literature on church music in the United States during the period from about 1960 to early 2004. The literature documents sweeping changes in religious groups, individual churches and their worship music, musical styles and types, and preferences of religious leaders, musicians, and the Christian public. Included and described are book-length monographs, bibliographies and collective biographies, selected dissertations, works lists, church and choral music journals, some journal articles, and World Wide Web sites relating to aspects of church music. Special emphasis is placed on studies relating to "tradition, change, and conflict" in American church music during the aforementioned four plus decades.

1. With few exceptions, titles chosen for inclusion were published or reprinted between 1960 and early 2004. A few landmark works that appeared earlier were, however, included.
2. Studies in languages other than English were not sought.
3. Selected general music reference works with significant information on American church music were included.
4. Reference resources to the journal literature of church music were included.
5. Many unpublished dissertations offer worthwhile studies of specific topics in American church music. As a result, approximately one hundred and twenty selected individual dissertations were annotated. Topics discussed in these studies are wide-ranging, for example, church music of specific religious and ethnic groups; of states and regions of the United States; genres, styles and type of church and sacred music; music ministry; contemporary music and worship; bibliographies of music and music literature.

2. ORGANIZATION

The guide is divided into ten main sections, with each subdivided into more specific headings:

I. *General Music Reference*
II. *Church Music Reference*
III. *Church Music in Periodicals*
IV. *Historical Studies*
V. *Regional Studies*
VI. *Religious and Ethnic Groups*
VII. *Church and Sacred Music Genres*
VIII. *Music Ministry*
IX. *Tradition, Change, and Conflict*
X. *Church and Worship Music Web Sites*

I

General Music Reference

DICTIONARIES AND ENCYCLOPEDIAS

1. Apel, Willi, ed. *Harvard Dictionary of Music.* 2nd ed., rev. and enlarged. Cambridge, Mass.: Belknap Press of Harvard University Press, 1969. xv, 935 p. SBN 674–37501–7 ML100 .A64 1969

   Music dictionary/encyclopedia. Some articles signed; brief bibliographies accompany longer articles. Musical examples and illustrations. Revised and expanded in 1986 as *The New Harvard Dictionary of Music,* edited by Don Michael Randel (item 6). Although Randel's edition updates scholarship, many articles were not carried over to the new edition.

2. District of Columbia Historical Records Survey. *Bio-Bibliographical Index of Musicians in the United States of America since Colonial Times.* Prefatory material by Charles Seeger, Harold Spivacke, and H. B. Dillard. Prepared under the supervision of Leonard Webster Ellinwood and Keyes Porter. 2nd ed. Washington, D.C.: Music Section, Pan American Union, 1956. xxiii, 439 p. ML106.U3 H6; reprints, New York: Da Capo Press, 1971. ISBN 0–306–70183–9 ML106.U3 H6 1971; New York: AMS Press, 1972. ISBN 0–404–08075–8 ML106.U3 H6 1972; St. Clair Shores, Mich.: Scholarly Press, 1972. ISBN 0–403–01362–3 ML106.U3 H6 1972b

   Includes approximately 10,000 individuals significant to the history of music in the United States before mid-twentieth century. For each entry, provides dates if known, brief description of the person's career (singer, conductor, etc.), and references to writings about the person.

3. Ewen, David. *American Composers: A Biographical Dictionary.* New York: G. P. Putnam's Sons, 1982. 793 p. ISBN 0–399–12626–0 ML390 .E815 1982; reprint, London: R. Hale, 1983. ISBN 0–7090–0692–6 ML390 .E815 1983

   A biographical dictionary of three hundred selected U.S. composers from approximately 1800 to 1980. Includes foreign-born composers that became U.S. citizens. Handy for concise information about composers of sacred music, as well as secular music. Includes lists of principal works and bibliographies of writings about the composers. Index of programmatic titles.

4. Hitchcock, H. Wiley, and Stanley Sadie, eds. *The New Grove Dictionary of American Music.* London: Macmillan Press; dist. in United States by Grove's Dictionaries of Music, New York, 1986 (rep. 1992). 4 vols. ISBN 0–943818–36–2 ML101.U6 N48 1986, ML101.U6 N48 1992

   Authoritative music dictionary/encyclopedia about American music and musicians. More than nine hundred contributors. Unfortunately not indexed; however, information about church music can be found in articles about specific genres, composers, and stylistic periods. Numerous illustrations, photos, musical examples, charts, and graphs; selective bibliographies, discographies, and lists of compositions at end of most articles.

5. Jones, F. O., ed. *A Handbook of American Music and Musicians: Containing Biographies of American Musicians and Histories of the Principal Musical Institutions, Firms and Societies.* Canaseraga, N.Y.: F. O. Jones, 1886. 182 p. ML106.U3 J7; reprint, New York: Da Capo Press, 1971. SBN 306–70163–4 ML106.U3 J7 1971

   Unchanged reprint of the first edition. Invaluable source of information about American music, musicians (both native and foreign born), and musical subjects up to 1886. Brief entries for personal names, performing groups, individual compositions and collections of music, musical societies, conservatories, publishers, periodicals, instrument manufacturers, music dictionaries, etc. Many entries not found in modern reference books. Few lists of compositions in biographical entries; few musical examples.

6. Randel, Don Michael, ed. *The New Harvard Dictionary of Music.* Cambridge, Mass.: Belknap Press of Harvard University Press, 1986. xxi, 942 p. ISBN 0–674–61525–5 ML100 .N485 1986

   Music dictionary/encyclopedia. Revision and expansion of the *Harvard Dictionary of Music* (2nd ed., 1969) edited by Willi Apel (item 1). Emphasis on the tradition of Western art music, but also includes coverage of non-Western and popular music and of musical instruments of all cultures. Some articles signed; brief bibliographies accompany longer articles. Musical examples and illustrations.

7. Sadie, Stanley, ed. *The New Grove Dictionary of Music and Musicians.* 2nd ed. London: Macmillan Reference; New York: Grove's Dictionaries, 2001. 29 vols. ISBN 0–333–60800–3 ML100 .N48 2001

   Exceptional resource. Predecessor: *Grove's Dictionary of Music and Musicians,* edited by Eric Blom. Authoritative music dictionary/encyclopedia. Universal in scope. Two appendixes: (1) index of terms used in articles on non-Western music, folk music, and kindred topics; and (2) list of approximately twenty-five hundred contributors. Information about church and worship music in the United States can be found in articles on specific genres, composers, and stylistic periods, among other topics. Numerous illustrations, photos, musical examples, charts, and graphs; selective bibliographies at end of most articles; some discographies. A new on-line version was launched in January 2001, available by subscription.

8. Slonimsky, Nicolas, ed. *Baker's Biographical Dictionary of Musicians.* Centennial ed. New York: Schirmer Books, 2001. 6 vols. ISBN 0–02–865525–7 ML105 .B16 2001

   An important biographical dictionary, with thousands of composer entries. Includes bibliographies and composer works lists.

9. Wilson, Charles Reagan, and William R. Ferris, eds. *Encyclopedia of Southern Culture.* Chapel Hill: University of North Carolina Press, 1989. xxi, 1634 p. ISBN 0–8078–1823–2 F209 .E53 1989

   An extensive, one-volume encyclopedia divided into twenty-five subject categories. The subject category "Music" consists of signed articles about musicians and secular and sacred music, including "Sacred Harp" (Harry Eskew), "Spirituals" (Dena J. Epstein), "Revival Songs" (David Warren Steel), and "Shape-Note Singing Schools" (Steel). Illustrations and photos; expansive index.

BIBLIOGRAPHIES OF MUSIC LITERATURE

General Works

10. *American Reference Books Annual.* 1st ed. Littleton, Colo.: Libraries Unlimited, 1970–. (Annual). ISSN 0065–9959 Z1035.1 .A55

    Excellent resource; reviews reference books in many subject disciplines, including music. Issued annually.

11. Duckles, Vincent H., and Ida Reed. *Music Reference and Research Materials: An Annotated Bibliography.* 5th ed. Michael A. Keller, advisory

editor; indexed by Linda Solow Blotner. New York: Schirmer Books, 1997. xviii, 812 p. ISBN 0–02–870821–0 ML113 .D83 1997

Invaluable reference resource. Includes more than four thousand selected, published music reference and research tools. Broadly organized into thirteen divisions, including: dictionaries and encyclopedias; histories and chronologies; guides to musicology; bibliographies of music literature; bibliographies of music; reference works on individual composers and their music; catalogs of music libraries and collections; catalogs of musical instrument collections; histories and bibliographies of music printing and publishing; discographies and related sources; yearbooks, directories, and guides; electronic information resources; and bibliography, the music business, and library science. Each division subdivided by subject. The expansive index is a valuable tool for locating items relating to church and worship music in the United States.

12. Heintze, James R. *American Music before 1865 in Print and on Records: A Biblio-Discography.* Preface by H. Wiley Hitchcock. Rev. ed. Brooklyn, N.Y.: Institute for Studies in American Music, Conservatory of Music, Brooklyn College of the City University of New York, 1990. xiii, 248 p. ISBN 0–914678–33–7 ML120.U5 H458 1990

Listing of pre-1865 music "in-print" at the time of publication of this resource. Based on collections of The American University, the Library of Congress, and the University of Maryland at College Park. Organized in three parts: (1) annotated bibliography of printed music; (2) music in facsimile reprints; and (3) discography. Printed music classified broadly by performing forces. Gives basic bibliographic information about each work along with annotative remarks. Discography arranged in alphabetical order by composer. Due to the organization of this resource, one would need to know the names of composers or titles of compositions to locate entries for church music; however, a number of works may be found in the "Choral" section. Two expansive indexes: (1) recording manufacturer index and (2) composers, compilers, and titles.

Music Literature and Periodical Indexes

13. *Christian Periodical Index: An Index to Subjects and Authors and to Book and Media Reviews.* Cedarville, Ohio: Association of Christian Librarians, 1956–. ISSN 0069–3871 Z7753 .C5

International in scope. Index of English language articles and reviews written from an evangelical perspective. Indexes approximately 120 religious periodicals. Currently published quarterly, cumulated annually and every three years. Consists of an author index, subject index, and index of reviews.

Many useful topics found in the subject index: choirs (music); choral conducting; choral conductors; choral singing; choruses, sacred; church music; conducting; conductors (music); music; music and youth; music festivals; music in churches; music publishers; music rehearsals; musical accompaniment; musicians; organ (musical instrument); and United States—church history. Also available as an electronic resource.

Christian Periodical Index. CD-ROM. Cedarville, Ohio: Association of Christian Librarians; New York: Silver Platter International N. V., 2002.

Currently covers titles from 1979 to the present, with ongoing retrospective indexing planned.

14. *Guide to Social Science and Religion in Periodical Literature.* Flint, Mich.; Clearwater, Fla.: National Periodical Library, 1972–. ISSN 0017–5307 (1970–1988), ISSN 1054–0946 (1989–) Z7753 .G83

Originally published as *Guide to Religious and Semi-Religious Periodicals* (National Library of Religious Periodicals, 1969–1971). Currently indexes about one hundred periodicals on social science and religion topics. Most periodicals published in the United States. Fruitful subjects for research include: church, music ministry; hymnology, hymns and songs of faith; music; spiritual life; and music and singing. See also entries for specific religious groups.

15. Hixon, Donald L. *Music in Early America: A Bibliography of Music in Evans.* Metuchen, N.J.: Scarecrow Press, 1970. xv, 607 p. ISBN 0–8108–0374–7 ML120.U5 H6

Index to published materials containing printed musical notation found in Charles Evans' *American Bibliography* and the Readex Corporation's microprint edition of *Early American Imprints, 1639–1800,* edited by C. K. Shipton. Excludes periodicals, newspapers, and other serial publications. Six sections: (1) alphabetical listing of composers/editors/compilers included in the *Early American Imprints* microprint edition and (2) alphabetical listing of composers/editors/compilers not included in the *Early American Imprints* microprint edition; (3) biographical sketches of composers; (4) composer/editor index; (5) title index; and (6) numerical index based on Evans serial numbers, with cross references to sections (1) and (2). Bibliography of sixteen writings.

16. *International Index to Music Periodicals: IIMP Full Text.* Alexandria, Virginia: Chadwyck-Healey. Computer file: http://iimpft.chadwyck.com/

Electronic information resource; available to subscribers. Indexes more than 415 international music periodicals from over 20 countries with

80-plus full-text titles. Also indexes music articles and obituaries appearing in the *New York Times* and the *Washington Post.*

17. *The Music Index: A Subject-Author Guide to Music Periodical Literature.* Warren, Mich.: Harmonie Park Press, 1949–. ISSN 0027–4348 ML118 .M84

Print edition published monthly since 1949; accompanied by annually-published complete subject heading list with cross references. First issue indexed 41 periodicals, all in English; now indexes more than 670 periodicals published in some twenty countries. Subject heading list and monthly issues incorporated into annual volumes. Subjects include forms and types of music, individuals, and music compositions. Cites reviews of books and dissertations, music, music recordings, and performances.

The Music Index Online. Warren Mich.: Harmonie Park Press, 1999–. URL: http://www.hppmusicindex.com/

Developed by Conway Greene Publishing Company. Cumulates entries from 1979 to 2001. Surveys over 670 international music periodicals.

The Music Index on CD-ROM. CD-ROM. Warren, Mich.: Harmonie Park Press, 2000. ISSN 1066–1514.

Cumulates entries from 1979 to 2000. Annual, cumulative updates planned.

18. *Religion Index One. Periodicals.* Chicago; Evanston, Illinois: American Theological Library Association, 1949–. (Semiannually). ISSN 0149–8428 Z7753 .A5

Index published semiannually with annual cumulative issue. International in scope. Comprised of a subject index, author and editor index, and scripture index. Useful topics found in subject index include: church music; contemporary Christian music; hymn tunes; hymns; liturgy; Masses; music concerts in churches; and music in worship. Available in part as an electronic resource published as *ATLA Religion Database on CD-ROM* (American Theological Library Association, 1983–).

Religion Index Two. Multi-Author Works. Chicago; Evanston, Illinois: American Theological Library Association, 1970–. (Annually). ISSN 0149–8436 Z7751.R35

Index published annually with quinquennial cumulations: separately published collections of writings by four or more authors on religious topics. International in scope. Comprised of subject index, author and editor index, and scripture index. For recommended topics found in subject index, see *Religion Index One. Periodicals* above. Available in part as an electronic resource published as *ATLA Religion Database on CD-ROM.*

19. *RILM Abstracts of Music Literature.* 1–. 1967–. New York: RILM International Center. (Quarterly). ISSN 0033–6955 ML1 .I83

    RILM Abstracts of Music Literature. Electronic information resource. New York: RILM, 1997–. URL: http://www.rilm.org

    Official journal of the International Repertory of Music Literature organization. Provides abstracts of music literature. Inclusion based on quality. International in scope. All abstracts in English; foreign titles translated. Published quarterly, each issue has an author index; fourth issue has an author, title, and subject index for the year. Also published as an electronic information resource. Mode of access: World Wide Web.

Dissertations and Master's Theses

20. Adkins, Cecil, and Alis Dickinson, eds. *International Index of Dissertations and Musicological Works in Progress.* 1st ed. Philadelphia: American Musicological Society, 1977. 422 p. ML128.M8 I5

    Adkins, Cecil, and Alis Dickinson, eds. *Doctoral Dissertations in Musicology.* 7th North American ed., 2nd International ed. Philadelphia: American Musicological Society; International Musicology Society, 1984. 545 p. ML128.M8 I5 1984

    Adkins, Cecil, and Alis Dickinson, eds. *Doctoral Dissertations in Musicology: February 1984–April 1995.* 2nd cumulative ed. Philadelphia: American Musicological Society, 1996. ix, 406 p.

    Doctoral Dissertations in Musicology-Online. Philadelphia: American Musicological Society; Bloomington, Indiana: School of Music, Indiana University, 1996–. Computer file: http://www.music.indiana.edu/ddm/

    Listing of doctoral dissertations in musicology, completed and "in progress." Entries classified by subject, including choral music, conducting, performance practice, sacred music, and so on. Project began in 1952 under the direction of Helen Hewitt, but charge since passed to Cecil Adkins in 1966, joined by Alis Dickinson in 1968. Supplements appeared in various publications, including *American Music Teacher, Acta musicologica, The Journal of the American Musicological Society,* and were issued privately to society members by the American Musicological Society. Cumulative editions have been published throughout the years with the last being the February 1984–April 1995 edition cited above. The print edition is no longer issued in favor of the *Doctoral Dissertations in Musicology-Online* database; Mode of access: World Wide Web.

21. *Dissertation Abstracts International: A, The Humanities and Social Sciences.* Ann Arbor, Mich.: University Microfilms International, 1969–. (Monthly). Z5053 .D57

Author-prepared abstracts of doctoral dissertations from educational institutions throughout the world. Published monthly accompanied by an annual cumulative author index.

Dissertation Abstracts Online. Ann Arbor, Mich.: ProQuest, 1986? Electronic information resource.

Electronic database available to subscribers. Access to 1.5 million doctoral dissertations and selected master's theses. Provides citations only for dissertations from 1861 to June, 1980; provides citations and abstracts for dissertations from July, 1980 to the present. Web version published as *ProQuest Digital Dissertation.*

22. Heintze, James R. *American Music Studies: A Classified Bibliography of Master's Theses.* Detroit, Mich.: Information Coordinators, 1984. xxv, 312 p. ISBN 0–89990–021–6 ML128.M8 H44 1984

Source for bibliographic control of master's theses "whose subject matter pertains to American music in an historical, sociological, or analytical manner." Almost twenty-four hundred entries. Few entries annotated. Excludes theses that examine instrumental or vocal programs in individual schools. Organized into seven subject categories: research and reference materials; historical studies (including composers, performance, secular music, sacred music, and education); theory; ethnomusicology; organology; special topics (including philosophy, communications, and libraries); and related fields (theater and dance). Sacred music section, part of the historical studies subject category, lists 267 writings and is further divided into narrower subject categories (general, vocal and instrumental, individuals, denominations). Three indexes: author, subject, and expansive geographic index.

DISCOGRAPHIES

See: Heintze, James R. *American Music before 1865 in Print and on Records: A Biblio-Discography* (item 12)

CHRONOLOGIES

23. Gleason, Harold, and Warren Becker. *Early American Music: Music in America from 1620 to 1920.* 2nd ed. Bloomington, Indiana: Frangipani Press, 1981. ix, 201 p. ISBN 0–89917–265–2 ML200 .G55 1981

Originally published under the title *American Music from 1620–1920* (University of Rochester, 1955). Series of fifteen outlines covering music in the United States from the time of early settlement at Plymouth and Jamestown through the end of World War I. Includes bibliographies and lists of music compositions for each outline; facsimiles and illustrations.

24. Hall, Charles J. *A Chronicle of American Music: 1700–1995.* Foreword by Gerard Schwarz. New York: Schirmer Books; London: Prentice Hall International, 1996. xi, 825 p. ISBN 0–02–860296-X ML200 .H15 1996

 Chronology of significant events, individuals, organizations, and publications in music history in the United States, 1700 to 1995, with correlation to significant highlights in U.S. art and literature and world culture. Extensive, expansive index.

II

Church Music Reference

DICTIONARIES AND ENCYCLOPEDIAS

General Works

25. Carroll, Joseph Robert. *Compendium of Liturgical Music Terms.* Toledo, Ohio: Gregorian Institute of America, 1964. 86 p. ML108 .C29

    Dictionary of liturgical music terms common to the Catholic Church.

26. Cross, F. L., ed. *The Oxford Dictionary of the Christian Church.* 3rd ed., rep. with corrections, ed. by E. A. Livingstone. New York: Oxford University Press, 1997. xxxvii, 1786 p. ISBN 0–19–211655-X BR95 .O8 1997

    Dictionary of subjects related to Christian churches, Catholic and Protestant. Universal in scope. Information on musical genres (anthem, hymns, Mass, etc.), significant individuals, groups, places, movements, and other topics. Greater emphasis on non-music topics. Cross-referenced. Short bibliographies follow some entries.

27. Davidson, James Robert. *A Dictionary of Protestant Church Music.* Metuchen, N.J.: Scarecrow Press, 1975. xvi, 349 p. ISBN 0–8108–0788–2 ML102.C5 D33

    Definitions of more than three hundred terms, some brief, others lengthy essays. All but briefest entries include a bibliography. Protestant church interpreted as "any Christian church other than those of the Roman Catholic . . . or Eastern traditions." Emphasis on Lutheran subjects. Unavoidably, includes terms commonly used in Catholic worship. A selective list of topics that relate to music in the United States include: anthem; camp meeting; congregational singing; "Dwight's Watts;" evangelistic music; folk song; fuging tune; gospel song; handbells; hymnody; hymn tune; lining-out; oratorio; psalmody; singing school; and spiritual. Cross-references between entries. Expansive index.

28. Foley, Edward, ed. *Worship Music: A Concise Dictionary.* Collegeville, Minn.: Liturgical Press, 2000. xx, 332 p. ISBN 0–8146–5889-X ML102.C5 W67 2000

Dictionary of terms, titles, and significant individuals relating to the religious and ritual aspects of music. Brief bibliographies provided for some entries.

29. Hughes, Anselm. *Liturgical Terms for Music Students: A Dictionary.* Boston, Mass.: McLaughlin & Reilly, 1940 (rep. 1972). iv, 40 p. ML108.H85 L5; St. Clair Shores, Mich.: Scholarly Press, 1972 (rep. 1977). iv, 40 p. ISBN 0–403–01363–1 ML108.H85 L5 1972

Pocket-sized dictionary of liturgical terms associated with music of the Catholic Church.

30. Julian, John, ed. *A Dictionary of Hymnology: Setting Forth the Origin and History of Christian Hymns of All Ages and Nations.* 2nd ed. London: John Murray, 1907 (rep. London: John Murray, 1915, 1925, 1957; New York: Scribner, 1915; New York: Dover Publications, 1957; New York: Gordon Press, 1979; Grand Rapids, Mich.: Kregel Publications, 1985). xviii, 1768 p. ISBN 0–8490–1719-X (1979), ISBN 0–8254–2960–9 (1985) BV305 .J8

First published in 1892; revised with supplement in 1907. Dictionary of hymns, composers, authors, translators, and other significant individuals and organizations commonly found in hymn books of English-speaking countries, primarily Great Britain and the United States. Also includes entries for subjects, such as hymnody, breviaries, missals, primers, psalters, and sequences. A 122-page supplement provides annotated list of principal hymnals issued since first edition and updates research source with additional information on authors, translators, and subjects. Tables; indexes to (1) first lines and (2) persons.

31. Poultney, David. *Dictionary of Western Church Music.* Chicago, Illinois: American Library Association, 1991. xxii, 234 p. ISBN 0–8389–0569–2 ML102.C5 P77 1991

Approximately four hundred terms, genres, and entries on significant composers associated with Catholic, Episcopal/Anglican, Lutheran, Methodist, and Baptist music practices. Only a few of the composers included are American. Many topics, however, deal with American music. Musical examples; list with addresses of American publishers of church music; list of societies and organizations associated with church music; bibliography of selected sacred music periodicals; index of compositions discussed and/or exemplified, arranged by composer.

32. Webber, Robert, ed. *Music and the Arts in Christian Worship.* Nashville, Tenn.: StarSong Publishing Group, 1994. 2 vols.: xliii, 828 p. ISBN 1–56233–014–4 (vol. 1), ISBN 1–56233–140-X (vol. 2) BV290 .M87 1994

Extensive encyclopedia; articles contributed by various authors. Most useful sections on church music are the first two: "Music and the Arts among the Churches," a discussion of the music of more than 50 churches, and "Music in Worship," which addresses theological issues and surveys singing in worship, congregational song, jubilation, use of musical instruments, planning music for worship, and music leadership. Remainder of volumes deal with visual arts, drama and dance, and the art of language. Musical examples and illustrations; expansive index; bibliographies of writings follow sections.

Biographical Dictionaries and Collective Biographies

33. Baxter, Clarice Howard, and Videt Polk. *Gospel Song Writers Biography.* Dallas, Tex.: Stamps-Baxter Music & Print, 1971. 306 p. ML390 .B37

Biographical entries for approximately one hundred American gospel song composers ranging from Lowell Mason (1792–1872) through the mid-twentieth century. Photos or illustrations of all composers.

34. Buehler, Kathleen D. *Heavenly Song: Stories of Church of God Song Writers and Their Songs.* Anderson, Indiana: Warner Press, 1993. x, 116 p. ISBN 0–87162–647–0 ML390 .B928 1993

Biographical information and discussion of the hymns of seventeen selected Church of God composers. Numerous photos and musical examples; separate indexes for composers and titles.

35. Burrage, Henry S. *Baptist Hymn Writers and Their Hymns.* Portland, Maine: Brown Thurston, 1888. xi, 682 p. BV380.A1 B9; abridged ed., Portland, Maine: Brown Thurston, 1888. xi, 174 p. BV380.A1 B92 1888; reprint (microfiche), Evanston, Ill.: American Theological Library Association, 1989.

Written more than a century ago, still a valuable source for biographical information on Baptist hymnodists, especially those lesser-known, and their hymn texts. International in scope; one 303-page section devoted to American Baptist hymnodists, mid-eighteenth century to late nineteenth century.

36. Claghorn, Charles Eugene. *Women Composers and Hymnists: A Concise Biographical Dictionary.* Metuchen, N.J.: Scarecrow Press, 1984. xiii, 272 p. ISBN 0–8108–1680–6 BV325 .C58 1984

    Biographical dictionary of 600 women composers and 155 women hymnists from Protestant, Catholic, and Jewish traditions. Chronologically, ranges from twelfth century to modern times; geographically, includes North America and Western Europe. More than one hundred hymnists/ composers from the United States. Provides composer name, dates, first line of well-known hymns, and brief biographical entry. Bibliography of twenty-six items; index by nationality.

37. Hall, Jacob Henry. *Biography of Gospel Song and Hymn Writers.* New York: Fleming H. Revell, 1914. 419 p. ML390 .H25

    Dated, but still useful. Biographical information on seventy-six selected gospel song and hymn writers arranged in chronological order from early eighteenth century to early twentieth century. Includes illustrations or photos of each.

38. Hatfield, Edwin F. *The Poets of the Church: A Series of Biographical Sketches of Hymn-Writers with Notes on Their Hymns.* New York: Anson D. F. Randolph, 1884. vii, 719 p. PR508.S3 H3; reprint, Boston, Mass.: Milford House, 1972. ISBN 0–87821–028–9 BV325 .H38 1972; reprint, Boston, Mass.: Longwood Press, 1977. ISBN 0–8934119–5–7 PR508.C65 H3 1977; reprint, Detroit, Mich.: Gale Research, 1978. ISBN 0–8103–4291-X PR508.C65 H3 1978

    Originally published in 1884 and reprinted several times. Biographical dictionary of nearly three hundred hymn writers who flourished primarily in the seventeenth through nineteenth centuries, although earlier writers are included. Not limited to U.S. hymn writers. Index of hymn titles.

39. Larson, Paul. *An American Musical Dynasty: A Biography of the Wolle Family of Bethlehem, Pennsylvania.* Bethlehem, Pa.: Lehigh University Press, 2002. 425 p. ISBN 0–934223–68–8 ML385 .L365 2002

    Biographies of Peter (1792–1871), Theodore Francis (1832–1885), and John Frederick Wolle (1863–1933), three influential Moravian Church composers and musicians. Photos; bibliography of compositions for each; glossary of terms; bibliography of about 130 writings; expansive index.

40. Metcalf, Frank J. *American Writers and Compilers of Sacred Music.* New York: Abington Press, 1925. 373 p. ML106.U3 M3; reprint, New York: Russell & Russell, 1967. ML106.U3 M3 1967

Dated, but useful. Chronologically arranged biographical dictionary of nearly one hundred American composers and compilers of sacred music. Also covers the Revivalist Group (1868–1872), camp meeting songs, and hymnody and psalmody in Washington, D.C. Facsimiles and photographs; expansive index.

41. Powell, Mark Allan. *Encyclopedia of Contemporary Christian Music.* Peabody, Mass.: Hendrickson, 2002. 1088 p. + 1 CD-ROM. ISBN 1–56563–679–1 ML102.C66 P68 2002

Biographical information for approximately two thousand artists associated with contemporary Christian music. CD-ROM included.

42. Robinson, Henrietta Fuller, and Carolyn Cordelia Williams. *Dedicated to Music: The Legacy of African American Church Musicians and Music Teachers in Southern New Jersey, 1915–1990.* Introduction by Clement Alexander Price. Cherry Hill, N.J.: Africana Homestead Legacy Publishers, 1996 (rep. 1997). xxi, 185 p. ISBN 0–9653308–0-X (1996), ISBN 0–9653308–4–2 (1997) ML385 .R69 1996, ML385 .R69 1997

Biographical dictionary of thirty-three church musicians and music teachers. Photographs and tables; bibliography of more than one hundred writings and a few scores, recordings, Web sites, and CD-ROMs; expansive index.

43. Rogal, Samuel J. *Guide to the Hymns and Tunes of American Methodism.* New York: Greenwood Press, 1986. xxii, 318 p. ISBN 0–313–25123–1 BV415.A1 R64 1986

Texts and tunes of Methodist hymnody, 1878 to 1966. Six sections: (1) alphabetical listing of poets followed by first lines of hymns and location within six major Methodist hymnals; (2) biographical information on the poets; (3) biographical information about composers and arrangers; (4) index to first lines of hymns; (5) composer and tune-source index; and (6) tune title index. Bibliography of thirty-nine writings.

44. Thomson, Ronald W. *Who's Who of Hymn Writers.* London: Epworth Press, 1967. 104 p. BV325 .T5

Biographical dictionary of more than two hundred significant hymnists. Many are British, but U.S. hymnists are included. Bibliography of fifteen writings.

45. Wilhoit, Melvin Ross. "A Guide to the Principal Authors and Composers of Gospel Song of the Nineteenth Century." D.M.A. dissertation. Louisville, Ky.: Southern Baptist Theological Seminary, 1982. v, 361 p.

Information on twenty American-born authors/composers of gospel song in the nineteenth century. For each, provides an average of ten pages of biographical information, covering lineage, early years, marriage and family, adult years, religious background and affiliations, and contributions to the gospel song movement. Also includes a list of hymnal locations of words and/or music and an annotated bibliography of writings about the authors/composers. General bibliography of approximately two hundred writings and thirty hymnals and song collections.

See also: Arnold, Corliss Richard. *Organ Literature: A Comprehensive Survey* (item 475); Clency, Cleveland Charles. "European Classical Influences in Modern Choral Settings of the African-American Spiritual: A Doctoral Essay" (item 490); Cross, F. L., ed. *The Oxford Dictionary of the Christian Church* (item 26); Edwards, George Thornton. *Music and Musicians of Maine: Being a History of the Progress of Music in the Territory Which Has Come to Be Known as the State of Maine, from 1604 to 1928* (item 204); Glover, Raymond F., ed. *The Hymnal 1982 Companion* (item 305); Hood, George. *A History of Music in New England: With Biographical Sketches of Reformers and Psalmists* (item 483); Julian, John, ed. *A Dictionary of Hymnology: Setting Forth the Origin and History of Christian Hymns of All Ages and Nations* (item 30); Poultney, David. *Dictionary of Western Church Music* (item 31)

BIBLIOGRAPHIES OF MUSIC LITERATURE

General Works

Church Music

46. Brunkow, Robert deV., ed. *Religion and Society in North America: An Annotated Bibliography.* Introduction by Martin E. Marty. Santa Barbara, Calif.: American Bibliographical Center; Clio Press, 1983. xi, 515 p. ISBN 0–87436–042–0 Z7831 .R44 1983

    Abstracts of 4,304 periodical articles on the history of religion in the United States and Canada since the seventeenth century. The section devoted to sacred music contains fifty-three entries. Expansive 179-page subject index useful to locate other entries on church music topics; author index.

47. Heaton, Charles Huddleston. "A Church Music Bibliography." *Music: The AGO & RCCO Magazine*, 3 (Jan. 1969): 38–39; 3 (Mar. 1969): 24–25; 3 (Apr. 1969): 52–53

    Dated classified bibliography of approximately four hundred writings. Items in English or English translations. Subject classifications: bells; choir and conducting; history; hymnal companions; hymnody; organ; periodicals; plainchant; and [performance] practice.

48. Hsieh, Fang-Lan. *An Annotated Bibliography of Church Music.* Assisted by Jason M. Runnels. Preface by Harry Eskew. Lewiston, N.Y.: E. Mellen Press, 2003. iv, 255 p. ISBN 0–7734–6580–4 ML128.C54 H75 2003

Bibliography of 1,177 writings, hymnals, collections of music, and Web sites in the field of church music. Organized into the following categories: dictionaries and encyclopedias; bibliographies; church music history; philosophy of church music; music and worship; music and ministry; hymnology; indexes; and periodicals. General index.

49. Miller, Patricia Ruch. "Reference Sources for the Church Musician." *The American Organist,* 24/5 (May 1990): 198–204

Annotated bibliography of fifty-three writings. International in scope. Many sources relate to music in the United States. Includes a number of diction books, singing manuals, and resources for selecting music.

50. Powell, Martha C. *A Selected Bibliography of Church Music and Music Reference Materials.* Assisted by Deborah C. Loftis. Louisville, Ky.: Southern Baptist Theological Seminary, 1977. v, 95, 9 p. ML128.C54 P6

Classified, annotated bibliography. Superseded in part by *Music Reference and Research Materials: An Annotated Bibliography* (1997), edited by Vincent H. Duckles, Ida Reed, and Michael A. Keller (item 11). Church music topics include sacred music periodical indexes, dictionaries, encyclopedias, bibliographies, histories, periodicals, and specific studies on gospel music, worship, choral music, organ music, hymns, individual religious groups, etc. Indexes: (1) authors and editors and (2) titles. Includes nine-page supplement.

51. Von Ende, Richard Chaffey. *Church Music: An International Bibliography.* Metuchen, N.J.: Scarecrow Press, 1980. xx, 453 p. ISBN 0–8108–1271–1 ML128.C54 V66

Not annotated. Literature on music in the Christian church, but also includes references to Jewish and Asian faiths. Includes 5,445 entries organized into 284 categories. International in scope, including citations in more than 25 different languages, though most are in English. Author, editor, and compiler index.

See also: Inoue, Tadashi. *Church Music Research Tools* (item 682)

U.S. Music

52. Heard, Priscilla S. *American Music, 1698–1800: An Annotated Bibliography.* Waco, Tex.: Baylor University Press, 1975. 246 p. ML120.U5 H4

Originated as the author's Master's thesis titled "An Annotated Bibliography of Music and References to Music Published in the United States before 1801" (Baylor University, 1969). The first chapter presents a brief survey of Charles Evans' *American Bibliography,* a publication that lists books, pamphlets, and periodicals published in the United States before 1800. The second chapter surveys the development of musical bibliography in the U.S. The main portion of the study is a 199-page annotated bibliography of entries that pertain to music drawn from Evans' *American Bibliography* and organized into three sections: entries which include music notation; entries pertaining to music; and entries not in microprint. For each entry, gives author, title, publication information, annotation, and references to other writings where the work is discussed. In addition to this extensive bibliography, the author provides a general bibliography of fifty-five writings on the subject; expansive index.

53. Heintze, James R. *Early American Music: A Research and Information Guide.* New York: Garland, 1990. xii, 511 p. ISBN 0–8240–4119–4 ML120.U5 H46 1990

Annotated bibliography of 1,959 writings on early American music from its beginning to 1820. Includes critical and facsimile editions of music. Chapter 6 presents several topical studies relating to church music: sacred music; psalmody and singing schools; hymnody; and music of ethnic and religious groups. Author/title index and subject index.

54. Horn, David. *The Literature of American Music in Books and Folk Music Collections: A Fully Annotated Bibliography.* Metuchen, N.J.: Scarecrow Press, 1977. xiv, 556 p. ISBN 0–8108–0996–6 ML120.U5 H7

Electronic version: Boulder, Colo.: NetLibrary, 1999. URL: http://www.netlibrary.com/summary.asp?ID=6613

Annotated bibliography of 1,388 writings about music and folk music collections published prior to 1976. A majority of the writings pertain to instrumental music and song; sections most likely to list studies on sacred music are: church music; hymnody; psalmody; religious folk song; spirituals; and gospel. As an appendix, 302 additional writings without annotation are listed. Expansive index. Also published as an e-book. Mode of access: World Wide Web.

Horn, David, and Richard Jackson. *The Literature of American Music in Books and Folk Music Collections: A Fully Annotated Bibliography: Supplement I.* Metuchen, N.J.: Scarecrow Press, 1988. xvi, 570 p. ISBN 0–8108–1997-X ML120.U5 H7 Suppl. 1

Electronic version: Boulder, Colo.: NetLibrary, 1999. URL: http://www.netlibrary.com/summary.asp?ID=6612

Supplement to Horn's *The Literature of American Music in Books and Folk Music Collections: A Fully Annotated Bibliography* (1977). Annotated bibliography of 996 writings about music and folk music collections published between 1975 and 1980, including titles omitted from the previous compilation. Similar organization to its earlier counterpart. Sections most likely to list studies on sacred music are: church music; hymnody; psalmody; religious folk song; spirituals; and gospel. As an appendix, 323 additional writings without annotation are listed. Separate indexes for subjects, names, and titles. Also published as an e-book. Mode of access: World Wide Web.

Marco, Guy A. *The Literature of American Music III, 1983–1992.* Lanham, Md.: Scarecrow Press, 1996. xviii, 451 p. ISBN 0–8108–3132–5 ML120.U5 M135 1996

Electronic version: Boulder, Colo.: NetLibrary, 1999. URL: http://www.netlibrary.com/summary.asp?ID=6614

Continuation of Horn's *The Literature of American Music in Books and Folk Music Collections: A Fully Annotated Bibliography* (1977) and its supplement (1988) by Horn and Jackson. Annotated bibliography of 1,302 writings about music and folk music collections published between 1983 and 1993. Sections most likely to list studies on sacred music are: history of American music; church music; and black music. Separate indexes for titles, subjects, and authors. Also published as an e-book. Mode of access: World Wide Web.

55. Jackson, Richard. *United States Music: Sources of Bibliography and Collective Biography.* Brooklyn, N.Y.: Institute for Studies in American Music, Dept. of Music, Brooklyn College of The City University of New York, 1976. Rev. ed. vii, 80 p. ISBN 0–914678–00–0 ML120.U5 J2 1976

Very selective, annotated bibliography of ninety-plus writings on various aspects of music in the United States as well as sources of collective biographical information. Separate, brief sections on African-American music and church music. Resources related to church music can also be found amongst reference works and historical and regional studies. Index of individual and corporate names.

56. Krummel, Donald William. *Bibliographical Handbook of American Music.* Basingstoke: Macmillan, 1987; Urbana: University of Illinois Press, 1987. 269 p. ISBN 0–333–44631–3 (Macmillan), ISBN 0–252–01450–2 (UIP) ML120.U5 K78 1987

Annotated bibliography of 760 resources in the area of American music, including bibliographies, works lists, discographies, indexes, and electronic information resources. Chapter 12, "Sacred Music," lists thirty-three historical bibliographies, denominational and other special lists, and analytical indexes. Other writings pertaining to church music periodicals, hymnology, Jewish music, and so on, may be located by using the general index.

57. Krummel, Donald William, Jean Geil, Doris J. Dyen, and Deane L. Root. *Resources of American Music History: A Directory of Source Materials from Colonial Times to World War II.* Urbana: University of Illinois Press, 1981. 463 p. ISBN 0–252–00828–6 ML120.U5 R47

Annotated bibliography of 1,689 resources of sacred and secular American music history before 1941. Surveys collections of manuscript and printed music, programs, catalogs, organizational and personal papers, iconography, and sound recordings (which includes piano rolls). Geographically limited to the United States, including Puerto Rico. Index.

Music Literature and Periodical Indexes

58. *Annotated Guide to Periodical Literature on Church Music.* Philadelphia: Music Article Guide, 1971–1972. 2 vols.: vi, 41 p.; vii, 40 p. ML128.C54 A6

Significant, signed articles on church music appearing in U.S. periodicals in 1971 and 1972, respectively. Articles drawn from issues of *Music Article Guide.* 1971 volume has thirty-four entries covering: organ; choirs and choral art; church music affairs in general; hymnology; bells and bell programs; and miscellaneous topics. 1972 volume has 211 entries covering: organ; choirs and choral art; hymnology; bells and bell programs; sacred music; and church music in general. Both volumes heavily cross-indexed.

59. *The Catholic Periodical and Literature Index.* Haverford, Pa.: Catholic Library Association, 1930–. ISSN 0008–8285 AI3 .C32

Formerly titled *Catholic Periodical Index.* Compiled indexes issued annually. Now indexes approximately 175 significant Catholic periodicals most published in the United States, but also in some twenty other countries and regions. Includes foreign-language periodicals. Divided into four sections: (1) subject index; (2) author and editor index; (3) book title index; and (4) book review index. Useful indexed subjects include: church music; Mass (music); and music–religious aspects.

Dissertations and Master's Theses

60. Hartley, Kenneth R. *Bibliography of Theses and Dissertations in Sacred Music.* Detroit, Mich.: Information Coordinators, 1967. viii, 127 p. ML128.S2 H4

Lists approximately fifteen hundred Master's theses and doctoral dissertations written in the United States dealing with aspects of sacred music. Organized alphabetically by state, then by educational institution. Provides author, title, degree earned, and date of completion. Not annotated. Several indexes: author, biographies, works of individual composers, subject, and stylistic period.

61. Porter, Thomas H. "Dissertations and Theses Related to American Hymnody, 1964–1978." *The Hymn,* 30/3 (July 1979): 199–204, 221

 Bibliography of eighty-five dissertations and theses in American hymnody. Provides Library of Congress microfilm numbers and *Dissertation Abstracts International* citations.

Special and Subject Bibliographies

African-American Music

62. Floyd, Samuel A., and Marsha J. Reisser. *Black Music in the United States: An Annotated Bibliography of Selected Reference and Research Materials.* Millwood, N.Y.: Kraus International Publications, 1983. xv, 234 p. ISBN 0–527–30164–7 ML128.B45 F6 1983

 Resources relating to African-American music in the United States. Includes writings on religious folk music, spirituals, and gospel music, as well as historical surveys on African-American worship music. Three indexes: titles; authors, editors, and compilers; and expansive general index.

63. Jackson, Irene V. *Afro-American Religious Music: A Bibliography and a Catalogue of Gospel Music.* Westport, Conn.: Greenwood Press, 1979. xiv, 210 p. ISBN 0–313–20560–4 ML128.S4 J3

 First part: classified bibliography of 873 writings in six sections: (1) African American: general history, culture, anthropology, and sociology; (2) ethnomusicology, dance, and folklore; (3) African and African-American folksongs; (4) religious folksongs: spirituals, hymns, blues, and gospels; (5) African-American church/African-American religion; and (6) Caribbean: religion, music, culture, folklore, and history. Second part: bibliography of gospel music published between 1937 and 1965, including works of almost five hundred composers. Entries supply composer, dates, titles of compositions, publisher, and sources of music. Two indexes: (1) expansive index to bibliography and (2) index to catalog.

Choral Music

64. DeVenney, David P. *American Choral Music since 1920: An Annotated Guide.* Berkeley, Calif.: Fallen Leaf Press, 1993. xviii, 278 p. ISBN 0–914913–28-X ML128.C48 D45 1993

Completes DeVenney's four-volume bibliographic survey of this repertory. Annotated bibliography of almost two thousand sacred and secular choral works by seventy-six composers active in the United States between 1920 and 1993, along with annotated bibliography of more than two hundred selected writings. Excludes most hymns, folk songs and spirituals, works written specifically for stage, choruses published separately from larger works, works written for an ensemble of solo voices rather than choral ensemble, and arrangements other than those made by the composer. Information, when applicable and known, includes composer, title, opus number, date of composition or copyright, author or source of text, duration, publisher, and location of manuscript. Entries in the bibliography of music are cross-referenced to writings in the bibliography of selected writings. Five indexes: titles, authors and sources of texts, performing forces, durations, and an expansive index to the bibliography of selected writings.

65. DeVenney, David P. *Early American Choral Music: An Annotated Guide.* Berkeley, Calif.: Fallen Leaf Press, 1988. xx, 149 p. ISBN 0–914913–09–3 ML128.V7 D43 1988

Guide to American choral music, 1670 to 1825. Three main sections: bibliography of music, bibliography of writings, and indexes. Bibliography of music contains known choral works by thirty-two composers, giving opus number, date of composition, performing forces, author of text, duration of work if over ten minutes, published editions, name of collection, and location of composer's manuscript. Bibliography of 127 writings. Six indexes: (1) compositions by genre (extended works for mixed voices; other works for mixed voices; and works for men's and women's voices); (2) sacred works; (3) secular works; (4) titles; (5) texts, authors, translators; and (6) an expansive index to the bibliography.

66. DeVenney, David P. *Nineteenth-Century American Choral Music.* Berkeley, Calif.: Fallen Leaf Press, 1987. xxi, 182 p. ISBN 0–914913–08–5 ML128.C48 D48 1987

Major research tool on nineteenth-century American choral music. Three main sections: bibliography of music, bibliography of writings, and indexes. Annotated bibliography of music contains almost thirteen hundred choral works by composers active in the United States, nineteenth century through World War I. Composers who primarily wrote popular music excluded. Also excluded: stage works, hymns, hymn tunes, melodies borrowed from folk tunes, arrangements of other composers' works, separately published choruses, choruses intended for didactic purposes, and songs with optional choral endings or refrains. Music entries include, if known, opus number, date of composition, performing forces, author of text, duration, published editions, and location of the manuscript. Lacks qualitative remarks. Musical entries cross referenced to articles in the bibliography. Annotated bibliography of more than 130 writings present historical, theoretical, and analytical examinations of the works or performance practice and interpretation of the composers' music. Ten indexes: five indexes to compositions organized by genre and voicing; sacred works; secular works; titles; authors and translators; and index to the bibliography of writings.

67. Orr, N. Lee, and W. Dan Hardin. *Choral Music in Nineteenth-Century America: A Guide to the Sources.* Lanham, Md.: Scarecrow Press, 1999. ix, 135 p. ISBN 0–8108–3664–5 ML128.C48 O77 1999

Selective bibliography of approximately nine hundred sources pertinent to choral music in the United States from the 1820s through the early years of the twentieth century. Many sources relate to sacred music. Sections that include sacred music are: Moravian music; shape-notes; African-American music; slave songs and spirituals; denominational music; hymns and gospel songs; and church histories, among others. Includes individual chapters on leading ninteenth-century composers, e.g., Lowell Mason, Amy Beach, John Knowles Paine, and so on. Index.

68. Sharp, Avery T., and James Michael Floyd. *Choral Music: A Research and Information Guide.* New York: Routledge, 2002. xiv, 318 p. ISBN 0–8240–5944–1 ML128.C48 S53 2002

Classified, annotated bibliography of 513 choral music resources (monographs, bibliographies, discographies, dissertations, music journals, electronic databases, and World Wide Web sites). Materials selected published between 1960 and 2000. An expansive, fifty-six-page subject index will identify many items that relate to church music in the United States. Author and title indexes.

69. Whitten, Lynn, ed. *A Classified, Annotated Bibliography of Articles Related to Choral Music in Five Major Periodicals through 1980.* Lawton, Okla.: American Choral Directors Association, 1982. xviii, 233 p. ML1500 .C63

Covers more than one thousand choral-related articles in the following periodicals: *The American Choral Review* (1958–1980), *Church Music* (1966–1980), *The Journal of the American Musicological Society* (1948–1980), *Music and Letters* (1920–1980), and *The Musical Quarterly* (1915–1980). Several sections pertinent to church music, including choral music genres, choral composers, and writings on various types of choral groups. Authors and translators index and subject index.

See also: Edwards, Randy. *Revealing Riches & Building Lives: Youth Choir Ministry in the New Millennium* (item 578); Garcia, William Burres.

"Church Music by Black Composers: A Bibliography of Choral Music" (item 86)

Episcopal/Anglican

70. Gardner, Elaine C. *A Resource Guide.* S.1.: Music Commission, Diocese of Western New York, 1986. 42 p. ML128.P7 G33 1986

Annotated bibliography of resources relating to church music in Episcopal/Anglican churches. Includes hymns and service music, organ music, lectionaries and resource books, books on special topics (buying an organ, employment issues, and copyright law), journals, recordings, and a listing of approximately 350 recommended anthems.

Folk Music

71. Miller, Terry E. *Folk Music in America: A Reference Guide.* New York: Garland, 1986. xx, 424 p. ISBN 0–8240–8935–9 ML128.F74 M5 1986

Annotated bibliography of resources on folk music in the United States. Organized into nine chapters with Chapter 6 offering 107 writings on American psalmody and hymnody, Chapter 7 offering 179 writings on the singing school and shape-note tradition, and Chapter 8 offering 85 writings on African-American religious music and gospel music. See also entries relating to ethnic traditions, such as Native American, Eskimo, and Hawaiian music. Author index and expansive subject index.

See also: Horn, David. *The Literature of American Music in Books and Folk Music Collections: A Fully Annotated Bibliography* (item 54)

Hymnody

72. Clark, Keith C. "A Bibliography of Handbooks to Hymnals: American, Canadian, and English." *The Hymn,* 30/3 (July 1979): 205–209; 30/4 (Oct. 1979): 269–272, 276; 31/1 (Jan. 1980): 41–47, 73–74; 31/2 (Apr. 1980): 120–126

Annotated bibliography of (1) current handbooks, 1927–1980, (2) supplementary guides, hymnals with biographical notes, concordances of hymnals, and indexes, 1927–1980, (3) early companions and annotated hymnals, 1845–1927, and (4) supplementary guides, 1773–1927. American, Canadian, and English resources interfiled. Facsimiles and photographs.

73. Clark, Keith C. *A Selective Bibliography for the Study of Hymns, 1980.* Springfield, Ohio: Hymn Society of America, 1980. 43 p. ML3270 .H9 no. 33

Revision of *A Short Bibliography for the Study of Hymns* (Hymn Society of America, 1964). Not annotated. Bibliography of approximately 650 writings relating to hymnody, psalmody, African-American religious music, carols, and church music in general. International in scope with many writings on U.S. topics.

74. Music, David W. *Christian Hymnody in Twentieth-Century Britain and America: An Annotated Bibliography.* Westport, Conn.: Greenwood Press, 2001. xiv, 205 p. ISBN 0–313–30903–5 ML128.H8 M87 2001

    Annotated bibliography of 1,190 writings about twentieth-century British and American hymnody. Introduced by twenty four-page historical overview of the subject. General index.

    See also: Rogal, Samuel J. "A Bibliographical Survey of American Hymnody 1640–1800" (item 103)

Masses

See: DeVenney, David P. *American Masses and Requiems: A Descriptive Guide* (item 115)

Mennonite

75. Springer, Nelson P., and A. J. Klassen. *Mennonite Bibliography, 1631–1961.* Introduction by Cornelius J. Dyck. Scottdale, Pa.: Herald Press, 1977. 2 vols.: 531 p.; 634 p. ISBN 0–8361–1208–3 Z7845.M4 S67, BX8121.2.M4 S67

    A bibliography of writings on the Mennonite Church, Brethren in Christ Church, and Hutterian Brethren. Not annotated. In two volumes: Vol. 1: international, Europe, Latin America, Asia, and Africa; Vol. 2: North America, with 12,555 entries. Author and subject indexes. Subject index identifies music-related items. Useful subject headings include: church music, hymns, music, music in churches, psalms (music), and sacred vocal music.

Psalmody

76. Warrington, James. *Short Titles of Books Relating to or Illustrating the History and Practice of Psalmody in the United States, 1620–1820.* Philadelphia: private printing, 1898. 96 p. ML120.U5 W2; reprint (microforms), New York: New York Public Library, n. d.; Evanston, Ill.: American Theological Library Association, 1990; reprints (books), Pittsburgh, Pa.: Clifford E. Barbour Library, 1970. ML120.U5 W2 1970; New York: Burt Franklin, 1971. SBN 8337–36906 ML120.U5 W2 1971

Lists more than one thousand tunebooks and writings on psalmody in the United States during two centuries. Arranged chronologically. Includes publications published as early as 1538.

See also: Clark, Keith C. *A Selective Bibliography for the Study of Hymns, 1980* (item 73)

BIBLIOGRAPHIES OF MUSIC AND MUSIC INDEXES

General Works

77. *The Christian Music Directories: Printed Music, 2000–2001.* Nick Wagner, editorial director. San Jose, Calif.: Resource Publications, 1999. 1531 p. ML128.C54 C57 2000–2001

    Previously published as *The Music Locator.* Lists printed music for use in worship services: modern and traditional compositions, Christian popular music, instrumental music associated with worship, and inspirational music. Each entry provides title, composer/arranger, year of publication, publishing information, publisher number, and some descriptive notes. Indexes for song titles, composers, songbook titles, and publisher number. Updated by printed supplements; also available on CD-ROM.

    The Christian Music Directories: Recorded Music, 2002. Nick Wagner, editorial director. San Jose, Calif.: Resource Publications, 2002. 1937 p. ML156.4.R4 R38 2002

    Previously published as *The Recording Locator.* Discographical companion to *The Christian Music Directories: Printed Music.* Six indexes: song title, artist, album title, video, publisher codes, and publisher names. Updated by printed supplements; also available on CD-ROM.

    The Christian Music Directories. Update. Nick Wagner, editorial director. San Jose, Calif.: Resource Publications, 1999. (three issues a year)

    Serial publication that updates *The Christian Music Directories: Printed Music* and *The Christian Music Directories: Recorded Music.*

    Christian Music Finder. Nick Wagner, editorial director. CD-ROM. San Jose, Calif.: Resource Publications, 1997–. (Quarterly)

    Available by subscription. Searchable database of *The Christian Music Directories: Printed Music* and *The Christian Music Directories: Recorded Music,* organized by title, artist, collection, and composer.

78. Holz, Ronald W. *A Reference Guide to Current Salvation Army Music Publications.* New York: Salvation Army Territorial Music Bureau, 1981. iii, 83 p. ISBN 0–89216–049–7

Needs to be updated. Annotated bibliography of more than 150 Salvation Army music publications in-print as of 1981. Organized into four sections: (1) music for congregational singing, (2) vocal music, (3) instrumental music, and (4) books, periodicals, theory, and timbrel methods. Each section is further divided into useful subsections based on performing forces.

79. Wenk, Arthur. *Musical Resources for the Revised Common Lectionary.* Metuchen, N.J.: Scarecrow Press, 1994. xvi, 614 p. ISBN 0–8108–2909–6 ML128.P7 W46 1994

Designed to assist ministers, organists, and choirmasters in selection of hymns, organ music, and choral music appropriate to the lessons appointed for worship. Five main sections: (1) music for the church year (A, B, C); (2) scriptural index of hymn texts; (3) index of hymn preludes, including a brief list of melodic incipits; (4) scriptural index of anthems; and (5) indexes of organ works and choral works cited.

See also: *emusicquest* (item 681)

Special and Subject Bibliographies and Music Indexes

Anthems

80. Balshaw, Paul. "A Selected Repertoire of Anthems and Motets for American Protestant Liturgical Churches." D.M.A. dissertation. New York: University of Rochester, 1963. iii, 116 p.

Annotated bibliography of approximately two hundred selected works. Listed in alphabetical order by composer, grouped under Sundays and Feasts of the church year. Provides composer's dates, date of composition, title, publisher, performing requirements, source of text, and analytical comments. Two indexes: composers and texts. Bibliography of fourteen writings.

Cantatas

81. Dox, Thurston J. *American Oratorios and Cantatas: A Catalog of Works Written in the United States from Colonial Times to 1985.* Metuchen, N.J.: Scarecrow Press, 1986. 2 vols.: 1330 p. ISBN 0–8108–1861–2 ML128.O45 D7 1986

Excellent annotated catalog of more than 3,450 oratorios and cantatas composed in the United States by more than one thousand composers. Spans more than two hundred years. Grouped into four categories: oratorio, choral cantata, ensemble cantata, and choral theater. Information given, if known, includes title, publishing information, date of composition, terminological designation (oratorio, choral cantata, ensemble cantata, or choral theater), required vocal and instrumental forces, characters designated in the work, source of text, approximate performing time, length in pages, number of parts, sections, acts, or movements, location of published score or manuscript, Online Computer Library Center (OCLC) number, date and place of first performance, and reviews of premiere or later performances. Title, author, and topical indexes.

82. Evans, Margaret R. *Sacred Cantatas: An Annotated Bibliography, 1960–1979.* Jefferson, N.C.: McFarland, 1982. xviii, 188 p. ISBN 0–89950–044–7 ML128.C15 E9 1982

Three main divisions: brief history of the sacred cantata, bibliography, and indexes. History subdivided into two parts, 1900–1960 and 1960–1979, limited to cantatas in England and America. Bibliography of more than four hundred sacred cantatas. Criteria for inclusion: work or edition written for SATB choir, with occasional sectional divisions and occasional soloists; English text or English translation; copyrighted after 1959 and before 1980; seven to thirty minutes duration; text appropriate for use in a Protestant worship service. Excludes settings of traditional Mass text and new editions of cantatas by J. S. Bach. Entries provide names of composer, titles of works, description of required forces, publisher, selling agent, place, publisher's number, copyright date, approximate performance time, availability of scores and parts, source of text, plan of the work, vocal ranges, and other descriptive notes; includes journal reviews, listings of works in periodicals and journals, and recordings; comments on stylistic characteristics and level of difficulty. Three indexes: title index; main index of composers, giving title, level of difficulty, appropriate church-year season, and soloists if required; special occasions index, which lists cantatas appropriate for specific days or occasions, excluding Christmas and Easter (included in the main index).

Chant

83. Bryden, John R., and David G. Hughes. *An Index of Gregorian Chant.* Cambridge, Mass.: Harvard University Press, 1969. 2 vols.: xvi, 456; v, 353 p. SBN 674–44875–8 ML102.C45 B8

The index "attempts to cover that portion of the chant that was in general use for a considerable period of time." Two volumes: (I) alphabetical order of textual incipits and (II) numerical order of melodic incipits. For each entry, provides mode, textual incipit, category to which the chant belongs, sources, first note of the chant, melodic incipit, and final.

Choral Music

84. Dupere, George Henry. "Sacred Choral Repertoire for Mixed Voices: A Recommended Listing." *Choral Journal,* 32/3 (Oct. 1991): 25–37

    Annotated list of approximately 280 choral works "appropriate for use in a variety of church situations." Criteria for selection: (1) SATB to SATB-divisi, (2) unaccompanied or keyboard accompaniment, and (3) no more than moderate difficulty. Provides title, composer, publisher, music number, subject (i.e., Christmas, general, praise, communion, etc.), and brief comments.

85. Eslinger, Gary S., and F. Mark Daugherty, eds. *Sacred Choral Music in Print.* 2nd ed. Philadelphia: Musicdata, 1985. 2 vols.: xiii, 1322 p. ISBN 0–88478–017–1 ML128.V7 E78 1985

    Updates *Choral Music in Print* (1974), edited by Thomas R. Nardone, James H. Nye, and Mark Resnick; *Choral Music in Print: 1976 Supplement,* edited by Thomas R. Nardone, *Sacred Choral Music in Print: 1981 Supplement,* edited by Nancy K. Nardone, and *Music in Print Annual Supplement* (1982–1984). Bibliography of all sacred choral music in print organized alphabetically by composer. International in scope, although works published in the United States, Canada, and Western Europe are emphasized. Each entry provides composer, title, voicing, accompaniment, remarks, and pricing information (U.S. dollars). Heavily cross-referenced. Directory of publishers with addresses, noting distributors. Separately published arranger index and master index (composer and title). Indexes and supplements:

    Sacred Choral Music in Print, Second Edition: Arranger Index. Philadelphia: Musicdata, 1987. iii, 137 p. ISBN 0–88478–019–8 ML128.V7 E7837 1987

    Simon, Susan H., ed. *Sacred Choral Music in Print: 1988 Supplement.* Philadelphia: Musicdata, 1988. xiii, 277 p. ISBN 0–88478–022–8 ML128.V7 E78 1985 Suppl.

    Daugherty, F. Mark, and Susan H. Simon, eds. *Sacred Choral Music in Print: 1992 Supplement.* Philadelphia: Musicdata, 1992. xiii, 304 p. ISBN 0–88478–029–5 ML128.V7 E78 1985 Suppl. 2

    Sacred Choral Music in Print: Master Index 1992. Philadelphia: Musicdata, 1992. vi, 413 p. ISBN 0–88478–030–9 ML128.V7 S2 1992

    Daugherty, F. Mark, ed. *Sacred Choral Music in Print: 1996 Supplement.* Philadelphia: Musicdata, 1992. xiii, 225 p. ISBN 0–88478–039–2 ML128.V7 E78 1985 Suppl. 3

Sacred Choral Music in Print: Master Index 1996. Philadelphia: Musicdata, 1996. vi, 451 p. ISBN 0–88478–040–6 ML128.V7 S2 1996

The Sacred Choral Sourcebook. Philadelphia: Musicdata, 1997. v, 467 p. ISBN 0–88478–046–5 ML128.V7 S23 1997

Derived from *Sacred Choral Music in Print* (1985) and its supplements (1988, 1992, 1996). Organized in two parts: composer listings and title listings. Information is minimal; provides composers, main titles of works, and publishers of editions. Publisher directory.

86. Garcia, William Burres. "Church Music by Black Composers: A Bibliography of Choral Music." *The Black Perspective in Music,* 2/2 (1974): 145–157

Bibliography of choral music for the church by forty-one African-American composers. Lists composer with dates, title, voicing/scoring, publication information, catalog number, pagination, and price. Includes a list of recommended books on African-American music and a list of publishers with addresses. Information dated.

87. Laster, James. *Catalogue of Choral Music Arranged in Biblical Order.* 2nd ed. Landham, Md.: Scarecrow Press, 1996. vi, 711 p. ISBN 0–8108–3071-X ML128.C54 L4 1996

Bibliography of more than seven thousand choral works. Divided into three main sections: Old Testament, the Apocrypha, and New Testament. A majority of the works are choral octavos. Extended works, with few exceptions, are excluded. Offers little in regard to Episcopal/Anglican Service music. Gives King James biblical reference: book, chapter, verse(s); additional scripture used, or author of paraphrased text, or translator; composer, arranger, or editor; title; voicing, solos, and accompaniment; and publication information. Cross-references texts from other biblical locations. Expansive composer and title indexes.

Laster, James. *Catalogue of Choral Music Arranged in Biblical Order: Supplement.* Landham, Md.: Scarecrow Press, 2002. vii, 107 p. ISBN 0–8108–4138-X ML128.C54 L4 1996 Suppl.

Supplement to the 2nd edition. Provides approximately thirteen hundred additional titles with emphasis on music published since 1995. Same format with indexes as 2nd edition.

88. White, Evelyn Davidson. *Choral Music by African American Composers: A Selected, Annotated Bibliography.* 2nd ed. London: Scarecrow Press, 1996. viii, 226 p. ISBN 0–8108–3037-X ML128.C48 W5 1996

Revision of the author's *Selected Bibliography of Published Choral Music by Black Composers* (Howard University, 1975). Annotated list of compositions by 102 African-American composers and arrangers. Many are sacred works. Entries are graded for difficulty. Information provided includes title, copyright date, number of pages, voicing and soloists, vocal ranges, range of difficulty, accompaniment (if applicable), publisher, and catalog number. Also includes selected listing of twenty-five collections of Negro spirituals and brief biographical sketches of eight-four composers. Three appendixes: selected source readings; selected discography, including a catalog of selected *Voice of America Recordings*; and addresses of publishers and composers. Title index.

89. Wolverton, Vance D. "Literature Forum: Choral Settings of Psalm Twenty-Three in English: An Annotated Bibliography." *Choral Journal,* 35/9 (Apr. 1995): 47–56; 35/10 (May 1995): 33–38

Annotated bibliography of English-language choral settings of Psalm 23. Two parts: (1) approximately seventy works for mixed voices, (2) nearly forty for unison, treble, and men's voices. Each entry provides composer/arranger, title, performing forces, publisher, music number, estimated price, level of difficulty (easy to very difficult), and descriptive remarks.

See also: DeVenney, David P. *American Choral Music since 1920: An Annotated Guide* (item 64); DeVenney, David P. *Early American Choral Music: An Annotated Guide* (item 65); DeVenney, David P. *Nineteenth-Century American Choral Music* (item 66); Edwards, Randy. *Revealing Riches & Building Lives: Youth Choir Ministry in the New Millennium* (item 578); Werning, Daniel J. *A Selected Source Index for Hymn and Chorale Tunes in Lutheran Worship Books* (item 111)

Contemporary Music

90. McLean, Terri Bocklund, and Rob Glover. *Choosing Contemporary Music: Seasonal, Topical, Lectionary Indexes.* Minneapolis, Minn.: Augsburg Fortress, 2000. 263 p. ISBN 0–8066–3874–5 ML128.C54 M35 2000

Church year calendar with approximately thirteen hundred recommended contemporary musical works. Four indexes: revised common lectionary, scripture references, seasons and topics, and titles. "Contemporary music" defined as "worship songs . . . in a wide variety of popular musical styles: folk, ballad, gospel, country, rock, blues, [etc.]"

Episcopal/Anglican

See: Gardner, Elaine C. *A Resource Guide* (item 70)

Fuging Tunes

91. Kroeger, Karl. *American Fuging-Tunes, 1770–1820: A Descriptive Catalog.* Westport, Conn.: Greenwood Press, 1994. xvi, 220 p. ISBN 0–313–29000–8 ML120.U5 K76 1994

Describes nearly 1,300 fuging tunes published during the late 1700s and early 1800s. Brief biographic information about each composer. Provides tune name, numerical incipit, structure, duration, number of fuges, fuge length, order of vocal entry, rhythmic profile, fuge ending and entry distance, date and collection for earliest printing of tune, poetic meter, first line of text, source of text, and annotative notes. Heavy use of abbreviations (abbreviations explained in prefatory pages). Separate indexes for tune names, structure, fuge order of entry, fuge rhythm, text coordination and fuge time interval, date of first publication, alphabetical listing of sources, chronological listing of sources, meter, first lines, poetic sources, geographic index of composers, and tunes with multiple fuges.

Gospel Music

See: Jackson, Irene V. *Afro-American Religious Music: A Bibliography and a Catalogue of Gospel Music* (item 63)

Hymns and Hymnals

92. Diehl, Katharine Smith. *Hymns and Tunes: An Index.* New York: Scarecrow Press, 1966. lv, 1185 p. BV305 .D5 1966

Index to seventy-eight hymnals, most used in the United States, others in Canada, England, and Scotland. Includes hymnals for the following groups: Amish, Baptist, Catholic, Congregational, Disciples of Christ, Episcopal/Anglican, Brethren, Jewish, Lutheran, Mennonite, Methodist, Missouri Synod, Moravian, Mormon, New Jerusalem, Presbyterian, Quaker, Reformed, Seventh-Day Adventists, Unitarian, and a few unaffiliated groups. Five sections: (1) first lines and variants, (2) authors with first lines, (3) tune names and variants, (4) composers with tune names, and (5) solfeggio index to the melodies. Prefaced by historical information about hymns and hymnals. Several appendixes, including annotated bibliography of hymnals indexed. Also, bibliography of six relevant reference books.

93. Ellinwood, Leonard Webster, ed. *Dictionary of American Hymnology: First-Line Index: A Project of the Hymn Society of America.* New York: University Music Editions, 1984. Microfilm. ML102 .H95

Published as 179 microfilm reels. Compilation of 1.2 million first-line citations from approximately 192,000 hymnals published between 1640 and 1978 in North and South America. Provides first lines, refrains, titles, original first-lines of translated hymns, authors, translators, date of publication, and citation for hymnal.

Ellinwood, Leonard Webster, and Elizabeth Lockwood. *Bibliography of American Hymnals: Compiled from the Files of the Dictionary of American Hymnology: A Project of the Hymn Society of America, Inc.* New York: University Music Editions, 1983. Microfiche. ML128 .H8

Companion to *Dictionary of American Hymnology.* Published as microfiche. Lists seventy-five hundred hymnals published in the United States. Provides title, imprint, year of publication, compiler, pagination, location of copy indexed, intended religious denomination, and name of indexer.

94. Hannum, Harold Byron. *Psalms and Hymns and Spiritual Songs: Brief Comments on Hymns from the Church Hymnal of Seventh-Day Adventists.* Arlington, Calif.: La Sierra College, 1959. 117 p. ML3086 .H3

Annotated bibliography of more than one hundred hymns arranged in alphabetical order by title. The hymns, though not clarified in the text, are most likely drawn from *The Church Hymnal: Official Hymnal of the Seventh-Day Adventist Church* (Review and Herald, 1941). Bibliography of nine recommended writings; index of authors and composers.

95. Hawn, C. Michael. "The Tie that Binds: A List of Ecumenical Hymns in English Language Hymnals Published in Canada and the United States since 1976." *The Hymn: A Journal of Congregational Song,* 48/3 (July 1997): 25–37

"List of hymns found to be in common in 40 English language hymnals published in Canada and the U.S. . . . between 1976 and 1996." Presents information in six tables: (1) alphabetical list of hymnals surveyed; (2) hymnals surveyed by date of publication and according to faith tradition; (3) ecumenical hymnody list in alphabetical order according to first line, with authors/translators; (4) ecumenical hymnody list in order of frequency of appearance, with most commonly used hymn tune(s); (5) alphabetical provisional list of ecumenical hymnody written in or translated into English after 1950, with most commonly used hymn tune(s); and (6) a provisional list of ecumenical hymnody written or translated into English since 1950 or not commonly used before 1950.

96. Hedger, Wayne L. *Baptist Hymnal Indices: Considering the 1991* Baptist Hymnal, *1975* Baptist Hymnal, *1956* Baptist Hymnal, *and 1940* Broadman Hymnal. Cleveland, Tenn.: Glory Press, 1993. 135 p. ISBN 0–9630656–0–2 ML128.H8 H4 1993

Organized into three parts: (1) separate indexes for each of the four Baptist hymnals listed in the title above; (2) separate indexes of hymns unique to each of the four hymnals; and (3) consolidated indexes of the combined hymnals and a separate index for hymns in all four hymnals.

97. Kroeger, Karl, and Marie Kroeger. *An Index to Anglo-American Psalmody in Modern Critical Editions.* Madison, Wisc.: A-R Editions, 2000. viii, 143 p. ISBN 0–89579–471–3 M2.3.U6 R4 v.40

Kroeger, Karl, and Marie Kroeger. *An Index to Anglo-American Psalmody in Modern Critical Editions.* Madison, Wisc.: A-R Editions; S.1.: Tempo Production, 2001. CD-ROM

Index to 2,087 early American psalm and hymn tunes, fuging tunes, set pieces, and anthems in modern critical editions. Organized by tune name. Provides composer, tune type, first line of text, text source, and subject of text. Five indexes: composer, tune type, first line, author and text source, and subject. Also issued as a handy CD-ROM.

98. Leaver, Robin A. "Hymnals, Hymnal Companions, and Collection Development." *Notes: Quarterly Journal of the Music Library Association,* 47/2 (Dec. 1990): 331–354

Discusses significant "bibliographical literature essential for hymnal collection development." Provides a partially annotated bibliography of hymnals, hymnal companions, and handbooks organized into three categories: North America, English-Speaking World outside North America, and Non-English-Language Traditions. The North America category contains more than 160 items arranged by denominations.

99. Lorenz, Ellen Jane. *Hymnbook Collections of North America.* 2nd ed. Fort Worth, Tex.: Hymn Society of America, 1987 (rep. 1988). 87 p. ML111 .V64 1987

Published under the author's name of Ellen Jane L. Porter. Annotated list of 407 collections of hymn books in North America, most in the United States. Includes collections held by institutions and private individuals. Provides name, address, phone number in some cases, and brief description of the collection. Several indexes: (1) collection owners, (2) titles of collections, (3) geographical index, and (4) subject index.

100. Mason, Henry Lowell. *Hymn-Tunes of Lowell Mason: A Bibliography.* Cambridge, Mass.: University Press, 1944. ix, 118 p. ML134 .M46; reprint, New York: AMS Press, 1976. ML134 .M46 1976

Bibliography of 1,697 hymn tunes by the American composer, Lowell Mason (1792–1872). For many, gives name of tune, source, and date of composition.

101. Ressler, Martin E. *An Annotated Bibliography of Mennonite Hymnals and Songbooks, 1742–1986.* Quarryville, Pa.: M. E. Ressler; Lancaster, Pa.: Lancaster Mennonite Historical Society, 1987. 117 p. Z7800 .R47 1987

An annotated bibliography of 152 hymnals and songbooks published in North America by the (Old) Mennonite Church, by and for the Old Order and Beachy Amish, by the Reformed Mennonite Church, and by the Church of God in Christ, Mennonites. In addition, provides an annotated bibliography of 41 music-related publications.

102. Robinson, Charles S. *Annotations upon Popular Hymns.* New York: Hunt & Easton; Cincinnati, Ohio: Cranston & Curtis; Cleveland: F. M. Barton, 1893. 581 p. BV312 .R6

Annotated bibliography of 1,215 hymns, the majority drawn from *Laudes Domini: A Selection of Spiritual Songs, Ancient and Modern* (Century, 1884) and *New Laudes Domini: A Selection of Spiritual Songs, Ancient and Modern* (Century, 1892). Provides text of hymns and supplies historical and biographical notes about the hymns and hymn writers. Some illustrations of hymn writers; separate indexes for authors and first lines.

103. Rogal, Samuel J. "A Bibliographical Survey of American Hymnody 1640–1800." *Bulletin of the New York Public Library,* 78 (1975): 231–252

An annotated bibliography of approximately 275 printed collections of music and writings on U.S. hymnody, 1640 to 1800. Five sections: (1) psalm, hymn, and anthem collections; (2) psalm and hymn collections for children; (3) single hymns and anthems; (4) musical collections; and (5) prose tracts on hymnody and sacred music. Entries provide compiler, title, printer/publisher, place of publication, dates of earliest known edition, and identifies libraries and private collections where items are housed. Facsimiles.

104. Rogal, Samuel J. *The Children's Jubilee: A Bibliographical Survey of Hymnals for Infants, Youth, and Sunday Schools Published in Britain and America, 1655–1900.* Westport, Conn.: Greenwood Press, 1983. xliv, 91 p. ISBN 0–313–23880–4 Z7800 .R63 1983

Prefaced by an essay on the history and development of children's hymnody, 1655–1900. First section, pp. 1–26: annotated bibliography of 301 hymnals for children published in the United States. Second section, pp. 27–76: annotated bibliography of 805 hymnals for children published in Great Britain. Four indexes: (1) sponsoring denominations, organizations, institutions, and societies; (2) sponsoring churches and schools; (3) authors, compilers, editors, and contributors; and (4) printers and publishers. Tables in introductory essay.

105. Showalter, Gracie I. *The Music Books of Ruebush & Kieffer, 1866–1942: A Bibliography.* Foreword by Ray O. Hummel, Jr. Richmond: Virginia State Library, 1975. xii, 40 p. ML120.U5 S54

Music publications by the Ruebush & Kieffer firm in Dayton, Virginia, a prolific publisher of "shaped note" hymnals. Facsimiles; index of titles.

106. Spencer, Donald Amos. *Hymn and Scripture Selection Guide: A Cross-Reference Tool for Worship Leaders.* Rev. and expanded. Grand Rapids, Mich.: Baker Book House, 1993. 315 p. ISBN 0–8010–8339–7 BV312 .S67 1993

Intended for "pastors, ministers of music, choir directors, and leaders of any church-related organization." Includes 380 hymns and twelve thousand related scriptural references. Inclusion based on examination of hymnals of ten major denominations and five widely used interdenominational hymnals. Two sections: (1) alphabetical listing of hymn titles and (2) listing of scripture passages in biblical order. Cross references between the two sections. Index of hymn titles.

107. Stillman, Amy Ku'uleialoha. *Hawaiian Hymns: An Index of Hawaiian-Language Protestant Hymnals.* S.1.: S.p., 1993. 2 vols.

Not available for examination. An index for "locating Hawaiian hymn texts in hymnals compiled in [Hawaii] between 1823 and 1972." Includes chronologically arranged bibliography of Hawaiian-language hymnals, a listing of the contents of each hymnal by first line of each hymn, and a master index of first lines. Held at the University of Hawaii at Manoa and the Bishop Museum Library, both in Honolulu.

108. Temperley, Nicholas. *The Hymn Tune Index: A Census of English-Language Hymn Tunes in Printed Sources from 1535 to 1820.* Assisted by Charles G. Manns and Joseph Herl. Oxford: Clarendon Press; New York: Oxford University Press, 1998. 4 vols.: xviii, 469 p.; xi, 587 p.; 785 p.; 797 p. ISBN 0–19–311150–0 ML128.H8 T46 1998

Multi-faceted research tool for early British and American psalm and hymn tunes. Four volumes: Vol. 1, introduction and sources; Vol. 2, tune indexes; Vols. 3 and 4, tune census. An historical and technical introduction surveys the publication and significance of hymn tunes and provides key definitions and clarifies the coverage of the research. A bibliography of sources lists extant sources included in the tune index. Various tune indexes provide searching by musical incipit, tune names, composer, unusual text meters, and text incipit, as well as concordances to three significant publications. The tune census inventories 17,424 tunes associated with English-language hymn tunes in printed sources, 1535 to 1820. Numerous indexes throughout: general, title, composer, chronological, geographical, and so on.

109. Wallace, Robin Knowles. "What Are We Teaching Our Children?: Hymnody and Children at the End of the Twentieth Century." *The Hymn: A Journal of Congregational Song,* 50/3 (Jul. 1999): 11–19

Develops a core list of 44 hymns drawn primarily from denominational hymnals; also includes some children's hymnals and a selected list of hymns for children that appeared in *The Hymn* in 1985. Tables.

110. Wasson, D. DeWitt. *Hymntune Index and Related Hymn Materials.* Foreword by Robin A. Leaver. Lanham, Md.: Scarecrow Press, 1998. 3 vols.: xv, 698 p.; 699–1632 p.; 1633–2612 p. ISBN 0–8108–3436–7 ML3186 .W2 1998

Wasson, D. DeWitt. *Hymntune Index and Related Hymn Materials.* Lanham, Md.: Scarecrow Press, 2001. CD-ROM. ISBN 0–8108–4144–4 ML3186 .W2 2001

Catalog of nearly thirty-four thousand chorales, hymns, psalms, canons, spiritual songs, and so on, from more than three hundred hymnals of Jewish, Catholic, and Protestant churches in the United States and other countries. Volumes 2 and 3 comprise the main catalog. Each hymntune title is listed in alphabetical order along with: earliest known date of composition or publication; solfeggio representation of the tune; sources, such as composer, book or hymnals where a tune first appeared, place and date of first publication, and arranger; variant title(s); catalog references referring to the "Index Key" where hymntune-related collections are listed; other pertinent notes. Volume 1 complements the catalog by providing a variety of useful indexes. Selected indexes include: (1) index of composers; (2) alphabetical listing of hymnals; (3) listing of hymnals by denomination; (4) index of hymntune sources; and (5) melodic index of hymntunes based on solfeggio. Also published as a CD-ROM.

111. Werning, Daniel J. *A Selected Source Index for Hymn and Chorale Tunes in Lutheran Worship Books.* Saint Louis, Mo.: Concordia, 1985. 234 p. ML128.H8 W47 1985

Index of tunes contained in *Lutheran Worship* (Concordia, 1986) and *Lutheran Book of Worship* (Augsburg, 1978) and selected tunes in *The Lutheran Hymnal* (Concordia, 1941) and *The Service Book and Hymnal* (Augsburg, 1958). Provides title of hymn or choral tune along with a bibliography of music publications based on hymn or chorale tune. Each bibliography is classified in 2 parts: "organ" (works intended for organ or keyboard only) and "choral" (individual octavos or works within choral collections). A list of music publishers and distributors appended.

112. Wolf, Edward C. "Lutheran Hymnody and Music Published in America 1700–1850: A Descriptive Bibliography." *Concordia Historical Institute Quarterly,* 50/4 (Win. 1977): 164–185

Annotated bibliography of more than forty hymnals and hymn collections without music and more than twenty chorale books, tunebooks, and other music "which outline development of American Lutheran hymnody and music to 1850."

See also: Rogal, Samuel J. *Guide to the Hymns and Tunes of American Methodism* (item 43)

Instrumental Music

113. Devol, John. *Brass Music for the Church: A Bibliography of Music Appropriate for Church Use—A Total of 1309 Works—Including Brass Parts for One Trumpet up to Twenty-Piece Brass Choir.* Plainview, N.Y.: H. Branch Publishing, 1974. ix, 102 p. ISBN 0–89869–134–6 ML128.S2 D49

Title self explanatory; however, also includes works that employ voice, chorus, narrator, guitar, harp, organ, piano, woodwinds, percussion, and strings with brass. Provides title, composer/arranger, publishing information, voicing/instrumentation, level of difficulty, and occasion.

114. Suggs, Julian S. *Instrumental Music for Churches: A Descriptive Listing.* Nashville, Tenn.: Sunday School Board of the Southern Baptist Convention, 1982. 142 p. ML128.I5 I5 1982

Needs to be updated. Annotated bibliography of instrumental music suitable for use in Protestant churches. Organized into seven categories: solo, duet, and trio of mixed instruments; woodwind ensemble; string ensemble; brass ensemble; mixed ensemble; standard, full, and symphonic band; and orchestra. Provides title, composer/arranger/compiler/editor, publication information, level of difficulty, duration, and other pertinent notes. Index of arrangers and composers.

Masses

115. DeVenney, David P. *American Masses and Requiems: A Descriptive Guide.* Berkeley, Calif.: Fallen Leaf Press, 1990. xvii, 210 p. ISBN 0–914913–14-X ML128.C2 D4 1990

Annotated bibliography of more than one thousand Masses, Requiems, and individual Mass movements by composers active in the United States between 1776 and 1990, with an annotated bibliography of almost 150 writings on American Masses and Requiems. Bibliography of music divided into two parts: (1) 28 significant American Masses and Requiems and (2) 935 other Masses and Requiems. As applicable, includes composer, title, date of composition, performing forces, duration, textual notes, movements, publishing information or location of manuscript, commissioning agent, and cross references to relevant citations in the bibliography of writings. Some reviews of compositions cited. Eight indexes; most are expansive. Separate indexes for men's voices, women's voices, children's voices, and unison voices. Also, indexes for Requiem settings, titles, non-liturgical texts and authors, and the bibliography of writings.

Moravian

116. Cumnock, Frances, ed. *Catalog of the Salem Congregation Music.* Chapel Hill: University of North Carolina Press, 1980. 682 p. ISBN 0–8078–1398–2 ML125.N67 C8

A catalog of music used by the Moravian religious community in Winston-Salem, North Carolina. Organized into three sections: (1) Salem Congregational music, (2) Sisters' Collection, and (3) Liturgies. Each entry provides composer, title, musical incipit, title page information, and descriptive notes. Prefaced by a historical introduction that traces the development of the collection from 1771 to the 1840s.

117. Rau, Albert George, and Hans T. David. *A Catalogue of Music by American Moravians, 1742–1842: From the Archives of the Moravian Church at Bethlehem, Pa.* Bethlehem, Pa.: Moravian Seminary and College for Women, 1938. 118, A-X p. ML120.U5 R14; reprint, New York: AMS Press, 1970. ISBN 0–404–07206–2 ML120.U5 R14 1970

More than a catalog; provides biographical information on ten U.S. Moravian composers active in the late eighteenth and early nineteenth centuries along with annotated catalog of their music from holdings of the Moravian Church at Bethlehem, Pennsylvania. Facsimiles and illustrations.

118. Steelman, Robert, ed. *Catalog of the Lititz Congregation Collection.* Chapel Hill: University of North Carolina Press, 1981. 488 p. ISBN 0–8078–1477–6 ML136.L6852 M77

A bibliography of the Lititz Congregation Collection held at The Moravian Archives in Bethlehem, Pennsylvania, which consists of music manuscripts used by the Moravian religious communities at Lititz and Warwick, Pennsylvania, during the eighteenth and nineteenth centuries. Provides composer; title; musical incipit; information quoted from wrappers, title pages, captions, etc.; tempo, key, and duration; inventory of parts or scoring; and other miscellaneous information. A few facsimiles; indexes for composers and titles.

Oratorios

See: Dox, Thurston J. *American Oratorios and Cantatas: A Catalog of Works Written in the United States from Colonial Times to 1985* (item 81)

Organ Music

119. Edson, Jean Slater. *Organ Preludes: An Index to Compositions on Hymn Tunes, Chorales, Plainsong Melodies, Gregorian Tunes and Carols.* Metuchen, N.J.: Scarecrow Press, 1970. 2 vols.: v, 1169 p. ISBN 0–8108–0287–2 ML128.O6 E4

International in scope. Extensive index of organ works based on carols, chorales, hymn tunes, and plainsong and Gregorian tunes. Two volumes. Volume 1 is a composer index. Lists composer/arranger, nationality, dates, and titles with publisher. Many U.S. composers/arrangers can be located by browsing the index. Volume 2 indexes titles and variant titles and provides composer/arranger and a thematic incipit for most entries. Heavily cross-referenced.

Edson, Jean Slater. *Organ Preludes: Supplement.* Metuchen, N.J.: Scarecrow Press, 1974. viii, 315 p. ISBN 0–8108–0663–0 ML128.O6 E4 Suppl.

Provides corrections and updates the earlier publication. Same format.

120. Frankel, Walter A., and Nancy K. Nardone, eds. *Organ Music in Print.* 2nd ed. Philadelphia: Musicdata, 1984. xiii, 354 p. ISBN 0–88478–015–5 ML128.O6 F7 1984

Updates *Organ Music in Print* (1975), edited by Thomas R. Nardone. Bibliography of all sacred and secular music for solo organ and organ with accompanying instruments or voice in print organized alphabetically by composer. Works in which the organ serves as accompaniment or continuo are excluded. International in scope, though works published in the United States, Canada, and Western Europe are emphasized. Each entry lists composer, title, instrumentation, publisher, remarks, and pricing information (U.S. dollars). Heavily cross-referenced. Directory of publishers with addresses, noting distributors. Separately published master index (composer and title). Indexes and supplements:

Daugherty, F. Mark, ed. *Organ Music in Print: 1990 Supplement.* Philadelphia: Musicdata, 1990. xiii, 297 p. ISBN 0–88478–026–0 ML128.O6 F7 1984 Suppl.

Cho, Robert W., Elisa T. Kahn-Ellis, Donald T. Reese, and Frank James Staneck, eds. *Organ Music in Print: 1997 Supplement.* Philadelphia: Musicdata, 1997. xiii, 210 p. ISBN 0–88478–043–0 ML128.O6 F7 1984 Suppl. 2

Organ Music in Print: Master Index 1997. Philadelphia: Musicdata, 1997. vi, 167 p. ISBN 0–88478–044–9 ML128.O6 F772 1997

121. Hunnicutt, Judy. *Index of Hymn Tune Accompaniments for Organ.* Fort Worth, Tex.: Hymn Society of America, 1988. 32 p. ML128.H8 H86 1988

Two parts: (1) annotated bibliography of 191 books and collections containing organ music based on hymn tunes; (2) alphabetical listing of hymn tune names, serving as an index to the bibliography and also providing page number in book or collection and musical key for each title.

122. Lawrence, Joy E. *The Organist's Shortcut to Service Music: A Guide to Finding Intonations, Organ Compositions and Free Accompaniments Based on Traditional Hymn Tunes.* 2nd ed. Cleveland: Ludwig, 1989. xxxii, 439, 9 p. ML128.O6 L36 1989

Guide for locating organ music based on hymn tunes found in hymnals of the following denominations: Baptist USA, Disciples of Christ, Episcopal/Anglican, Lutheran, United Church of Christ, Catholic, United Methodist, and United Presbyterian. Identifies 2,300 hymn tunes, 2,400 organ compositions, and 750 free accompaniments.

See also: Werning, Daniel J. *A Selected Source Index for Hymn and Chorale Tunes in Lutheran Worship Books* (item 111)

Piano Music

123. Vickers, Laura Maxey. "An Annotated List of Standard Piano Literature for Use in the Christian Church Service." D.M.A. dissertation. Norman: University of Oklahoma, 1996. vii, 328 p.

Topic introduced by a discussion of the evolution of church music and the role of the piano in American Christian church services. Followed by annotated bibliography of two hundred mostly-secular piano works deemed appropriate for Christian worship services. Repertoire from baroque musical period to the present. For each title, provides composer, description of music, level of difficulty (intermediate, upper intermediate, lower advanced, or advanced), occasion for which composition is suitable, historical era, key, meter, tempo, duration, and selected editions. Three indexes: (1) level of difficulty, (2) by occasion, and (3) by historical era of composition. Bibliography of approximately 90 writings and 110 editions of music.

Spirituals

124. Abromeit, Kathleen A. *An Index to African-American Spirituals for the Solo Voice.* Foreword by François S. Clemmons. Westport, Conn.: Greenwood Press, 1999. xiii, 199 p. ISBN 0–313–30577–3 ML128.S4 A27 1999

Four indexes to access approximately eighteen hundred African-American spirituals for solo voice, with or without accompaniment. Indexes: title index; first line index; alternate title index; and topical index (i.e., admonition/judgment; aspiration; Christmas; church; etc.). The title index supplies the location of the spiritual in published sources and a description of the music (i.e., text, melody, and piano accompaniment; text and chord symbols; text only; etc.).

Tunebooks

125. Britton, Allen Perdue, Irving Lowens, and Richard Crawford. *American Sacred Music Imprints, 1698–1810: A Bibliography.* Worcester, Mass.: American Antiquarian Society, 1990. xvi, 798 p. ISBN 0–912296–95-X ML128.H8 B68 1989

Extensive, annotated bibliography of sacred music, mostly tunebooks, printed in the British-American colonies and the United States, 1698 to 1810. For each main item, provides transcription of title page. For each specific item, gives pagination, page size, method of printing, date of publication, and other relevant notes. When known, identifies engraved plates, notes about the music, attributions to composers and sources, first printings, origin of the music (United States or elsewhere), number of "core repertory" pieces (101 sacred compositions most printed in America, 1698–1810) in each book, citations of works found in important bibliographies, location of copies, and additional notes, including cross-references, as warranted. Describes contents of each according to its main divisions. Includes compilers' prefatory statements (often quoted in part), which includes compilers' advertisements, endorsements, forewords, introductions, and prefaces. Five appendixes: (1) chronological list of imprints; (2) sacred sheet music, 1790–1810; (3) list of composers and sources; (4) the "core repertory;" and (5) geographical directory of engravers, printers, publishers, and booksellers. Two expansive indexes: (1) prefatory statements and (2) general index.

126. Metcalf, Frank J. *American Psalmody: Or, Titles of Books Containing Tunes Printed in America from 1721 to 1820.* New York: Heartman, 1917. 54 p. ML120.U5 M3; reprint, New York: Da Capo Press, 1968. x, 54 p. ML120.U5 M3 1968

Bibliography with library locations of more than two hundred books of early sacred music published in the United States. Updated in part and supplemented by Charles Evans' *American Bibliography* (1903), Allen P. Britton's "Theoretical Introductions in American Tunebooks to 1800" (University of Michigan, 1949), and Ralph R. Shaw and Richard H. Shoemaker's *American Bibliography: A Preliminary Checklist* (1958–1966). Harry Eskew contributed a new introduction to the 1968 reprint. Facsimiles.

127. Stanislaw, Richard J. *A Checklist of Four-Shape Shape-Note Tunebooks.* Institute for Studies in American Music, Department of Music, School of Performing Arts, Brooklyn College of The City University of New York, 1978. ix, 61 p. ISBN 0–914678–10–8 ML128.H8 S7

Originated from the author's doctoral dissertation, "Choral Performance Practice in the Four-Shape Literature of American Frontier Singing Schools" (University of Illinois at Urbana-Champaign, 1976). Annotated bibliography of 305 tunebooks utilizing four-shape shape-note notation published in the United States between 1798 and 1860. Organized alphabetically by compiler/author. Provides title, publication information, physical description, and location of tunebook when known. Bibliography of 57 writings about tunebooks; chronological listing of four-shape tunebooks; title index.

See also: Warrington, James. *Short Titles of Books Relating to or Illustrating the History and Practice of Psalmody in the United States, 1620–1820* (item 76)

Vocal Music

128. Espina, Noni. *Vocal Solos for Christian Churches: A Descriptive Reference of Solo Music for the Church Year Including a Bibliographical Supplement of Choral Works.* 3rd ed. Metuchen, N.J.: Scarecrow Press, 1984. xiii, 241 p. ISBN 0–8108–1730–6 ML128.V7 E8 1984

Revision of the second edition titled *Vocal Solos for Protestant Services* (1974). Primarily lists solo vocal music; also includes a selected bibliography of almost 150 oratorios, Masses, passions, cantatas, and extended anthems scored for solo voices and choir with or without accompaniment. International representation of composers from the sixteenth through twentieth centuries, with more than one hundred composers from the United States. Separate sections on "Negro Spirituals" and "Traditional Sacred Songs." Gives composer, title, scoring, language of text, and publisher. Three indexes: occasions, music for solo voices, and titles.

129. Laster, James, and Diana Reed Strommen. 2nd ed. *Catalogue of Vocal Solos and Duets Arranged in Biblical Order.* Lanham, Md.: Scarecrow Press, 2003. ix, 225 p. ISBN 0–8108–4838–4 ML128.S3 L38 2003

Bibliography of approximately forty-five hundred vocal solos and duets. Divided into three main sections: Old Testament, the Apocrypha, and New Testament. Gives King James biblical reference: book, chapter, verse(s); additional scripture used, or author of paraphrased text, or translator; composer, arranger, or editor; title; language; range; accompaniment; and publication information. Cross-references texts from other biblical locations. Expansive composer and title indexes.

130. Luther, David Alan. "An Annotated Compendium of Selected Vocal Solo Literature for Protestant Churches, 1958–1988." D.M.A. dissertation. Baton Rouge: Louisiana State University and Agricultural and Mechanical College, 1990. iii, 112 p.

Unavailable for examination. Annotated bibliography of 148 songs appropriate for Protestant worship services. Excludes gospel, contemporary Christian songs, and music more fitting for recitals. Gives title, composer, author of text, publisher, date, range, tessitura, occasion, accompaniment, and descriptive notes.

DISCOGRAPHIES

131. Kavanaugh, Patrick. *The Music of Angels: A Listener's Guide to Sacred Music from Chant to Christian Rock.* Foreword by Dave Brubeck. Chicago: Loyola Press, 1999. xiv, 334 p. ISBN 0–8294–1019–8 ML3000 .K44 1999

Concise survey of sacred music over the last two thousand years. Identifies important musicians, influential works, and musical styles. Each chapter provides a list of recommended recordings. Sections relating to U.S. church music include those on congregational music, hymnology, gospel music, and contemporary Christian music. Photographs and illustrations; bibliography of more than 450 writings; expansive index.

See also: Hall, Roger L. *A Guide to Shaker Music: With Music Supplement* (item 378); Moore, Berkley L. *Recordings Index* (item 716); *The Christian Music Directories: Printed Music, 2000–2001* (item 77)

TEXTS AND TRANSLATIONS

132. Bausano, William. *Sacred Latin Texts and English Translations for the Choral Conductor and Church Musician: Propers of the Mass.* Westport, Conn.: Greenwood Press, 1998. 278 p. ISBN 0–313–30636–2 ML54.8 .S23 1998

Offers poetic English translations of more than nine hundred Latin propers of the Mass for the entire church year as used in the Catholic Mass prior to Vatican Council II (1962–1965). Provides cross-referencing and listings of propers for various feasts and seasons of the year.

133. Jeffers, Ron. *Translations and Annotations of Choral Repertoire. Vol. 1, Sacred Latin Texts.* Corvallis, Ore.: Earthsongs, 1988. 279 p. ISBN 0–9621532–0–6, ISBN 0–9621532–1–4 (pbk.)

Literal word-by-word English translation of more than one hundred Latin texts, including the Roman and Requiem Mass, with prose rendering to restore proper word order and clarify meaning within appropriate liturgical context. Several additional features: description of liturgical year and Hours of Divine Office; glossary of terms; Latin pronunciation guide, which includes Austro-German variants; and selected settings of Latin texts with cross references of settings of texts by composers, styles, and voicings. Index of titles and first lines.

III

Church Music in Periodicals

134. *The American Organist.* 1–. 1967–. New York: American Guild of Organists. (Monthly)

     News of American Guild of Organists (AGO) and Royal Canadian College of Organists (RCCO) members, events, competitions, and so on; interviews; articles.

135. *Call to Worship: Liturgy, Music, Preaching and the Arts.* 35–. 2001–. Louisville, Ky.: Office of Theology & Worship. (Quarterly)

     Reformed Liturgy & Music. 1–34. 1981–2000. Louisville, Ky.: Joint Office of Worship of the United Presbyterian Church, U.S.A., and the Presbyterian Church, U.S., in cooperation with the Presbyterian Association of Musicians. (Quarterly)

     Formerly *Reformed Liturgy & Music.* A quarterly journal for planners of worship, pastors, educators, and musicians in the Reformed Presbyterian tradition. Offers articles, book reviews, worship aids, theological reflection, and other resources.

136. *CCM Magazine: Faith in the Spotlight.* 1–. 1986–. Nashville, Tenn.: Salem. (Monthly)

     Magazine about contemporary Christian music sponsored by CCM Magazine.com (item 713). Consists of feature articles, news, and reviews of music, books, and tours.

137. *The Choral Journal.* 1–. 1959–. Lawton, Okla.: American Choral Directors Association. (10 issues a year)

The Texas Choirmaster. 1–5. 1959–1964. Abilene, Tex.: Texas Choral Directors Association. (Quarterly)

Association news and events; three to four substantive articles per issue; reviews of recordings and books; twenty-five-plus reviews of choral works in each issue; research reports; bibliographies and annotated bibliographies a regular feature. Absorbed *The Texas Choirmaster*, Aug./Sept. 1964.

Paine, Gordon. *The Choral Journal: An Index to Volumes 1–18.* Lawton, Okla.: American Choral Directors Association, 1978. xv, 170 p. ML1.C6563 P3

Index to volumes 1–18 of *The Choral Journal.* Provides abstracts unless content is obvious from the title. Includes: major articles; "Short Subjects;" regular columns; book reviews and dissertation abstracts; and record reviews after September, 1973. Excludes: most American Choral Directors Association organizational news and announcements; articles featuring individual choirs and tours by individual choirs; columns whose content does not lend itself to indexing; record reviews before September, 1973 and all reviews of noncommercial recordings; and "Choral Review." Consists of two parts: (1) subject index (classified into 74 subjects) and (2) an expansive general index. General index is cross-referenced with subject index.

138. *The Chorister.* 48–. 1996–. Garland, Tex.: Choristers Guild. (10 issues a year)

Choristers Guild Letters. 1–47. 1949–1996. Garland, Tex.: Choristers Guild. (10 issues a year)

Journal of the Choristers Guild (item 711), available through membership in the Choristers Guild. Formerly issued as *Choristers Guild Letters.* Purpose: "Choristers Guild, a Christian organization, enables leaders to nurture the spiritual and musical growth of children and youth."

139. *Church Music: An Annual Publication of Church Music in America.* 1–[15]. 1966–1980. Saint Louis, Mo.: Concordia Publishing House. (Annually)

Journal produced by Concordia Teachers College, River Forest, Illinois, concerning music in the Lutheran church. Consists of feature articles and reviews of books and music.

140. *Church Musician Today: A Resource of Music and Worship Leaders.* 1–. 1997–. Nashville, Tenn.: LifeWay Christian Resources of the Southern Baptist Convention. (Monthly)

The Church Musician. 1–48. 1950–1997. Nashville, Tenn.: Sunday School Board of the Southern Baptist Convention. (Quarterly)

Journal of the Sunday School Board of the Southern Baptist Convention. Formerly published as *The Church Musician.* Consists of feature articles, book reviews, and worship resources (i.e., dramas, leadership tips, and Christian Web sites).

141. *Creator: The Bimonthly Magazine of Balanced Music Ministries.* 1–. 1978–. Healdsburg, Calif.: Creator Magazine. (Bimonthly)

http://www.creatormagazine.com/

Independent church music journal. Published in print and online versions. The print edition includes columns not yet integrated into the online version. Online version allows browsing of articles, editorials, clip art, Web links, publishers and catalogs, and so on.

142. *The Hymn: A Journal of Congregational Song.* 1–. 1949–. Fort Worth, Tex.: Hymn Society of the United States and Canada. (Quarterly)

Journal of The Hymn Society of the United States and Canada (item 722). Contains articles, reviews of books and music, and organizational news.

143. *Ministry & Liturgy.* 26–. 1999–. San Jose, Calif.: Resource Publications. (10 issues a year)

Modern Liturgy. San Jose, Calif.: Resource Publications. 1–25. 1974–1998. (8–10 issues a year)

http://www.rpinet.com/ml/index.html

Independent Catholic journal. Formerly published as *Modern Liturgy.* Described as a "professional magazine for members of the ministry team." Print subscriptions available. Online version offers liturgical music discussion group, among others. "Christian Music Finder" software available.

144. *Pastoral Music.* 1–. 1976–. Silver Spring, Md.: National Association of Pastoral Musicians. (Bimonthly)

Journal of the National Association of Pastoral Musicians (NPM) (item 665). Consists of feature articles, book and music reviews, and news relating to Catholic church music.

145. *Reformed Worship: Resources for Planning and Leading Worship.* 1–. 1986–. Grand Rapids, Mich.: CRC Publications. (Quarterly)

http://www.reformedworship.org/

Journal of the Christian Reformed Church, consisting of articles, news, and reviews. Supports traditional, contemporary, and blended worship services. Content broad enough to be helpful to different denominations. Published in print format; Web site complements print issue and offers additional assistance to worship planners. Online Classic Content area includes audio "Songs for the Season."

146. *Sacred Music.* 92–. 1965–. Saint Paul, Minn.: Church Music Association of America. (Quarterly)

Caecilia. 1–91. 1874–1965. Various publishers, including Omaha, Neb.: Society of Saint Caecilia. (Irregular)

http://www.musicasacra.com/sacredmusic.html

Oldest continuously published music journal of any kind in the United States. Continuation of *Caecilia* (1874–1965). Church Music Association of America (CMAA) is an organization for Catholic Church musicians and others. Journal consists of articles, news, and reviews of music, books, and recordings.

147. *Southern Baptist Church Music Journal.* 1–. 1984–. Louisville, Ky.: Southern Baptist Church Music Conference. (Annual)

Articles on Southern Baptist church music. Reviews of music and books.

148. *The Stanza.* 1–. 1977–. Springfield, Mo.: Hymn Society of America. (Semi-annual)

Newsletter of The Hymn Society of America. Forum for opinions and information requests. Announcements of contests, programs, lectures, and hymn festivals.

149. *The Tracker: Journal of the Organ Historical Society.* 1–. 1956–. Richmond, Va.: Organ Historical Society. (Quarterly)

Emphasizes American organ topics of the eighteenth, nineteenth, and twentieth centuries, with occasional articles on European topics.

See also: *CCMusic: Contemporary Christian Music* (item 714); *The Church Music Report* (TCMR) (item 679)

IV

Historical Studies

150. Chase, Gilbert. *America's Music: From the Pilgrims to the Present.* 3rd ed. Foreword by Richard Crawford. Discographical essay by William Brooks. Urbana: University of Illinois Press, 1987 (rep. 1992). xxiv, 712 p. ISBN 0–252–00454-X (1987), ISBN 0–252–06275–2 (1992) ML200 .C5 1987, ML200 .C5 1992

    Survey of sacred and secular music in the United States from the seventeenth century to the 1980s. Most useful sections concerning religious music include "The Musical Puritans," "Conflict and Reform," "Dissenters and Minority Sects," "American Pioneers," "Progress, Profit, and Uplift," "Fasola Folk," "Revival Hymns and Spiritual Songs," and "The Negro Spirituals." Musical examples, facsimiles, illustrations, and photographs; bibliography of more than seven hundred writings; discographical essay by William Brooks discusses recordings of concert, folk, commercial, and jazz music; general index favors the inclusion of persons and titles with less emphasis on subjects.

151. Covey, Cyclone. "Religion and Music in Colonial America." Ph.D. dissertation. Calif.: Stanford University, 1949. ix, 335 p.

    Music in relationship to religion in colonial United States, focusing on music of the following groups: Calvanistic, Puritan, Presbyterian, Baptist, Reformed Dutch, Reformed German, French Huguenot, Episcopal, Catholic, Lutheran, Pietists, Quaker, Wissahickon Mystic, Ephrata Cloister, Amish, Moravian Brethren, Methodist, General Baptist, Shakers, and

Jewish. Bibliography of nearly seventy-five scores and music collections and more than three hundred writings.

152. Crawford, Richard. *America's Musical Life: A History.* New York: W. W. Norton, 2001. xv, 976 p. ISBN 0–393–04810–1 ML200 .C69 2001

Chronological study. A few chapters devoted to sacred music. Includes musical examples, facsimiles, illustrations, and photographs; bibliography of nearly six hundred writings; expansive index.

153. Crawford, Richard. *An Introduction to America's Music.* New York: W. W. Norton, 2001. xiv, 555 p. ISBN 0–393–97409-X ML200 .C72 2001

Survey of sacred and secular music in the United States from Colonial times to the present. Musical examples, facsimiles, illustrations, and photographs; no bibliography; expansive index.

154. Davis, Ronald L. *A History of Music in American Life.* Malabar, Fla.: R. E. Krieger, 1980–1983. 3 vols.: xv, 301 p.; xv, 268 p.; xvii, 444 p. ISBN 0–89874–002–9 (vol. 1), ISBN 0–89874–003–7 (vol. 2), ISBN 0–89874–004–5 (vol. 3) ML200 .D3

Sacred and secular music in the United States. Three volumes: Vol. 1, "The Formative Years, 1620–1865;" Vol. 2, "The Gilded Years, 1865–1920;" and Vol. 3, "The Modern Era, 1920—Present." Early chapters of first volume provides most information on church and worship music. Topics include the music of the New England Puritans and Middle Colonies and the music of Francis Hopkinson (1737–1791), William Billings (1746–1800), and contemporaries. Rest of study primarily treats secular subjects. Bibliographies of approximately thirty to forty writings for each chapter; expansive indexes for each volume.

155. Dean, Talmage W. *A Survey of Twentieth Century Protestant Church Music in America.* Nashville, Tenn.: Broadman Press, 1988. 284 p. ISBN 0–8054–6813–7 ML3111.5 .D4 1988

Surveys twentieth-century Protestant music in the United States within the context of religious, social, economic, and political influences. Bibliography of more than one hundred writings; index.

156. Dickinson, Edward. *Music in the History of the Western Church: With an Introduction on Religious Music among the Primitive and Ancient Peoples.* New York: Scribner, 1902 (rep. 1903, 1916, 1923, 1925, 1927, 1928, 1931, 1953). viii, 426 p. ML3000 .D65; London: Smith, Elder, 1902. ML3000 .D65 1902b; New York: Greenwood Press, 1969. ISBN 0–8371–1062–9 ML3000 .D65 1969; New York: Haskell House Publishers, 1969. ISBN 0–8383–0301–3 ML3000 .D65 1969b; St. Clair Shores, Mich.: Scholarly

Press, 1970. ML3000 .D65 1970; New York: AMS Press, 1970. ISBN 0–404–02127–1 ML3000 .D65 1970b

Last two chapters, respectively titled “Congregational Song in England and America” and “Problems of Church Music in America,” present a late-nineteenth-century perspective on church music in the United States. Former chapter intertwines English and American church music, covering Puritan music, Calvinistic psalmody, hymnody, the contributions of Isaac Watts and that of Charles and John Wesley, to name only a few subjects. Problems of church music in America addressed in the latter chapter is stated: “Every form of church music known in Europe flourishes in America, but there is no American school of religious music.” Certainly dated. Of historical interest. Bibliography of nearly seventy writings; expansive index.

157. Edwards, Arthur C., and W. Thomas Marrocco. *Music in the United States.* Dubuque, Iowa: W. C. Brown, 1968 (rep. 1969). xi, 179 p. ML200 .E25

Early chapters treat church and worship music in the United States. Topics include psalters and the music of the Puritans, the singing school movement, psalms and hymns, religious folk songs, and the worship music of the Moravians, Seventh-Day Baptists, Mennonites, Pietists, Shakers, Catholics, Jews, and Mormons. Musical examples and facsimiles; bibliography of about 275 writings and an outdated discography of about three hundred recordings.

158. Ellinwood, Leonard Webster. *The History of American Church Music.* Rev. ed. New York: Da Capo Press, 1970. xiv, 274 p. ISBN 0–306–71233–4 ML200 .E4 1970

Corrected reprint of the 1953 edition. American church music from Spanish colonization (1494) to the mid-twentieth century. Topics include metrical psalmody, hymnody, fuging tunes, singing schools and early choirs, quartet choirs, boy’s choirs, significant composers and musicians, and choral repertory. Classified list of approximately four hundred church works and brief biographies of approximately seventy American church musicians appended. Photographs, illustrations, facsimiles, and musical examples; documented with end notes; expansive index.

159. Ellsworth, Donald Paul. *Christian Music in Contemporary Witness: Historical Antecedents and Contemporary Practices.* Grand Rapids, Mich.: Baker Book House, 1979. 229 p. ISBN 0–8010–3338–1 ML3001 .E9

Two parts: (1) “Music for Church Outreach: Historical Evidence” and (2) “Witness Music since 1960.” Part 1 surveys witness music from the early church and the middle ages to 1960; part 2 addresses witness music in the 1960s and 1970s. Discussion is broad but includes U.S. church music topics. Bibliography of nearly 250 writings; expansive index.

160. Ellsworth, Donald Paul. "Music in the Church for Purposes of Evangelism: Historical Antecedents and Contemporary Practices." D.M.A. dissertation. Los Angeles: University of Southern California, 1977. vii, 291 p.

Study organized into two parts: (1) historical evidence of music for church outreach land (2) witness music since 1960. Bibliography of extended works (thirty-seven nonseasonal musicals, eight seasonal musicals, ten children's folk musicals, ten music dramas, three oratorios, three services, and ten Masses) and thirteen folk hymnals.

161. Elson, Louis Charles. *The History of American Music.* Rev. by Arthur Elson. New York: Macmillan, 1925. xiii, 423 p. ML200 .E49 1925; reprint, New York: Burt Franklin, 1971. ISBN 0–8337–1055–9 ML200 .E46 1971

Originally published in 1904; revised in 1925 with additional material provided by Arthur Elson. Certainly dated, but may be of interest to compare the authors' historical perspective to more recent studies. Sections relating to church and worship music are: Chapter 1, "The Religious Beginnings of American Music," Chapter 2, "Early Musical Organizations," and Chapter 13, "Organists, Choir and Chorus Leaders." Illustrations, photos, musical examples, and facsimiles; bibliography of seventy-one writings; expansive index.

162. Etherington, Charles L. *Protestant Worship Music: Its History and Practice.* New York: Holt, Rinehart and Winston, 1962. x, 278 p. ML3100 .E8; reprint, Westport, Conn.: Greenwood Press, 1978. ISBN 0–313–20024–6 ML3100 .E8 1978

The most useful section relating to U.S. music is a chapter titled "Worship Music in the American Colonies." Additional coverage may be found in latter chapters, namely "The Century of Neglect (1750–1850)," "The Late Nineteenth Century," and "The Present State of Worship Music." Musical examples; various lists of music compositions; general bibliography of more than eighty items; expansive index.

163. Faulkner, Quentin. *Wiser than Despair: The Evolution of Ideas in the Relationship of Music and the Christian Church.* Westport, Conn.: Greenwood Press, 1996. xix, 251 p. ISBN 0–313–29645–6 BV290 .F28 1996

Examines church music practices from the early church to the late twentieth century. Concentrates on European church music; slight coverage of U.S. church music, included only briefly in the last chapter on the twentieth century. Musical examples and illustrations; bibliography of approximately 260 writings and thirteen music scores; index.

164. Gould, Nathaniel D. *History of Church Music in America: Treating of Its Peculiarities at Different Periods, Its Legitimate Use and Its Abuse, with Criticisms, Cursory Remarks and Notices Relating to Composers, Teachers, Schools, Choirs, Societies, Conventions, Books, etc.* Boston, Mass.: Gould and Lincoln, 1853. xii, 240 p. ML2911 .G69; reprint, New York: AMS Press, 1972. ISBN 0–404–02888–8 ML2911 .G69 1972

 Also published under the title *Church Music in America.* Account of sacred music in the United States from 1770 to 1850. Covers psalmody, contributions of William Billings and contemporaries, singing schools, music theory, music in churches, musical instruments, and music societies, academies, and conventions. Bibliography of seventy-six collections of sacred music for schools and churches published in the United States between 1810 and 1852.

165. Hamm, Charles. *Music in the New World.* New York: W. W. Norton, 1983. xiv, 722 p. ISBN 0–393–95193–6 ML200 .H17 1983

 Historical study of music of the United States. Chapters treating church and worship music include: Chapter 1, "The Music of the Native American," Chapter 2, "Psalms, Hymns, and Spiritual Songs in the Colonies," Chapter 6, "William Billings and Lowell Mason: Birth and Reform," and Chapter 10, "Shape-Note, Camp-Meeting, and Gospel Hymnody." Illustrations, photographs, musical examples, and facsimiles; classified general bibliography of more than 550 writings and discography of nearly two hundred recordings; expansive index.

166. Heintze, James R., ed. *American Musical Life in Context and Practice to 1865.* New York: Garland, 1994. x, 366 p. ISBN 0–8153–0816–7 ML200.4 .A4 1994

 Collection of ten articles about American musical life prior to 1865. Of special interest are: "Peter Erben and America's First Lutheran Tunebook in English" (E. C. Wolf); "The Anthem in Southern Four-Shape Shape-Note Tunebooks, 1816–1860" (D. W. Music); and "Catholic Church Music in the Midwest before the Civil War: The Firm of W. C. Peters & Sons" (R. D. Wetzel). Index of names and titles.

167. Hitchcock, H. Wiley. *Music in the United States: A Historical Introduction.* 4th ed. Final chapter by Kyle Gann. Upper Saddle River, N.J.: Prentice Hall, 2000. xviii, 413 p. ISBN 0–13–907643–3 ML200 .H58 2000

 U.S. music history from early settlements to the mid-1980s. Concise discussion of church music topics, as well as secular topics. Musical examples; bibliographies included at the end of most sections; expansive index.

168. Hooper, William L. *Church Music in Transition.* Foreword by Loren R. Williams. Nashville, Tenn.: Broadman Press, 1963. vi, 208 p. ML3111 .H66

History of church music with emphasis on hymnody of evangelical denominations in the United States, namely Baptists, Methodists, Presbyterians, and Disciples of Christ. Musical examples and facsimiles; bibliography of about 160 writings; expansive general index and expansive index of hymnals.

169. Howard, John Tasker. *Our American Music: A Comprehensive History from 1620 to the Present.* 4th ed. Supplementary chapters by James Lyons. Rev. bibliography by Karl Kroeger. New York: Crowell, 1965. xxii, 944 p. ML200 .H8 1965

History of music in the United States. Study is divided into three parts, with first part pertaining more closely with church and worship music. Topics covered include New England psalmody, seventeenth-century church organs, and contributions of Francis Hopkinson, James Lyon, William Billings, and Lowell Mason. A later chapter, Chapter 11, "Latter-Century and Present-Day Religious Music," covers the contribution of Dudley Buck, along with discussion of folk hymns, gospel song, and twentieth-century religious music. Classified bibliography of approximately one thousand writings; index.

170. Kaufmann, Helen Loeb. *From Jehovah to Jazz: Music in America from Psalmody to the Present Day.* New York: Dodd, Mead, 1937. xiii, 303 p. ML200.K21 F93; reprint, Freeport, N.Y.: Books for Libraries Press, 1968. ML200 .K23 1968; reprint, Port Washington, N.Y.: Kennikat Press, 1969. ISBN 0–8046–0565–3 ML200 .K23 1969

Survey of music in the United States. Chapters pertaining to church and worship music include: Chapter 1, "One Little Psalm Book and How It Grew," and Chapter 2, "Black Americans, Their Spirituals and Folk Songs." Remainder of book is devoted to secular music. Illustrations; documented with endnotes.

171. Kinscella, Hazel Gertrude. *History Sings: Backgrounds of American Music.* Rev. by Emile H. Serposs. Lincoln, Neb.: University Publishing, 1970. xi, 428 p. ML200 .K56 1970

First published in 1940 with two reprints (1948, 1957); revised in 1970. Survey organized chronologically, primarily within geographical groupings. Among topics covered are: Puritan song, *The Bay Psalm Book* (1640), singing schools, William Billings, the Old Swedes' Church, music in Bethlehem, Pennsylvania, Lowell Mason, spirituals, Native American music, California mission music, St. Olaf Lutheran choir, and the Mormon Tabernacle Choir. Musical examples, illustrations, and photos; index.

172. Liemohn, Edwin. *The Singing Church.* Columbus, Ohio: Wartburg Press, 1959. vi, 122 p. ML3000 .L53 1959

Traces the role of congregational singing from Old Testament to present times. Chapter 7, "New England Psalmody," covers the Ainsworth Psalter, the decline of psalmody, *The Bay Psalm Book* (1640), "regular singing," singing schools, the transition to hymnody, and prominent early composers in the United States. Chapter 9 is very brief and primarily concerns the organ in eighteenth-century American churches. Musical examples and facsimiles; classified bibliography of eighty-seven writings; index.

173. Lowens, Irving. *Music and Musicians in Early America.* New York: W. W. Norton, 1964. 328 p. ML200 .L7

Topics discussed include *The Bay Psalm Book* (1640), John Tufts' *Introduction to the Singing of Psalm-Tunes* (1721–1744), the shape-note tunebook *The Easy Instructor* (1798–1831), John Wyeth's *Repository of Sacred Music, Part Second* (1813), the origins of the American fuging-tune, the church song, and the contributions of Daniel Read (1757–1836), Lewis Edson (1748–1820), and Lewis Edson, Jr. (1771–1845). Music facsimiles and tables; appendixes include the complete text of Tufts' *Introduction to the Singing of Psalm-Tunes* and a bibliography of editions and issues of *The Easy Instructor*; index.

174. Lutkin, Peter Christian. *Music in the Church.* Milwaukee, Wis.: The Young Churchman Company, 1910. xii, 274 p. ML3001 .L8; reprint, New York: AMS Press, 1970. ISBN 0–404–04069–1 ML3001 .L95 1970

Historical survey. Organized into six chapters: hymn tunes; congregational singing; the organ; the organist and choirmaster; the vested male choir; and development of music in the Episcopal/Anglican Church. Information about American church music sprinkled throughout. Bibliography of nearly one hundred writings; index of tunes and expansive general index.

175. Mathews, W. S. B., ed. *A Hundred Years of Music in America: An Account of Musical Effort in America.* Chicago: G. L. Howe, 1889. ix, 715 p. ML200 .H85; reprint, Philadelphia: Theodore Presser, 1900. ML200 .M429 1900; reprint, New York: AMS Press, 1970. ISBN 0–404–04259–7 ML200 .M37 1970

Music in America from 1620 to the late nineteenth century. Early chapters, Chapters 1–4, have the most to offer on church and worship music. Subjects include psalmody, *The Bay Psalm Book* (1640), reaction against florid church music, William Billings, Lowell Mason, and so on. Chapter 10 examines psalmody after the Civil War. Chapter 13 covers organists and liturgical music. Nearly three hundred illustrations and photographs of significant individuals provide additional value. Includes supplementary biographical dictionary of approximately 275 American musicians. No bibliography; no index.

176. McCue, George, ed. *Music in American Society, 1776–1976: From Puritan Hymn to Synthesizer.* New Brunswick, N.J.: Transaction Books, 1977. 201 p. ISBN 0–87855–209-X, ISBN 0–87855–634–7 (pbk.) ML200.1 .M9

Twelve essays on music in the United States. Essays on church music or sacred music: "Social and Moral Music: The Hymn" (A. B. Caswell) surveys American hymnody; "The Other Side of Black Music" (K. B. Billups) examines African-American church music to some extent, but a majority of the essay examines nonreligious African-American music; and "The Music of American Indians" (C. J. Frisbie) discusses religious music of Native Americans. Facsimiles; documented with endnotes; expansive index.

177. Music, David W., ed. *We'll Shout and Sing Hosanna: Essays on Church Music in Honor of William J. Reynolds.* Fort Worth, Tex.: School of Church Music, Southwestern Baptist Theological Seminary, 1998. vi, 283 p. ML3100.M86 W4 1998

Fourteen essays. Those relating to church music in the United States: "William J. Reynolds" (J. S. Moore) presents biographical information about Reynolds; "An Apostolic Norm for Congregational Song" (B. H. Leafblad) traces biblical practice of the use of congregational song; "Current Theological Trends Affecting Congregational Song" (P. Westermeyer) examines theological and aesthetic imperative of quality for the congregational song; "And the Beat Goes On: The Continuing Influence of Popular Culture on Congregational Song" (M. Price) examines wide-spread influence of popular culture upon congregational song in the United States; "Worship and Learning Styles: Practical Applications for Worshipers and Worship Leaders" (R. Bradley) looks at learning styles in relationship to worship; "Barlow and Beyond" (M. L. VanDyke) discusses the contributions of Joel Barlow (1754–1812) to psalmody in America; "Sweet Chants that Led My Steps Abroad: Anglican Chant in Nineteenth-Century American Baptist Hymnals" (P. A. Richardson) studies the presence of Anglican chant in some mid- to late-nineteenth-century Baptist hymnals published in the United States; "Parody in Nineteenth- and Twentieth-Century American Hymnody" (P. G. Hammond) examines the use of *contrafactum* parody; "Sacred Harp Singing, Remnant of Frontier Religion" (C. R. Young) discusses shape-note singing practice and the repertory of the *Sacred Harp* (1844); "The Birth of a Classic: Sankey's 'The Ninety and Nine'" (M. R. Wilhoit) examines theories regarding the popular song; "From Civil War

Song to Children's Hymn: Jesus Loves the Little Children" (H. Eskew) demonstrates *contrafactum* parody of a Civil War song to create the children's hymn; and "William J. Reynolds: A Bibliography" (D.M. Music) is a classified bibliography of Reynolds's 100-plus writings and his more than 150 original compositions, arrangements, and editions of music.

178. Porter, Susan L., and John Michael Graziano, eds. *Vistas of American Music: Essays and Compositions in Honor of William K. Kearns.* Warren, Mich.: Harmonie Park Press, 1999. xii, 379 p. ISBN 0–89990–088–7 ML200.1 .V57 1999

Twenty-seven essays and musical compositions in honor of Kearns. Two essays relate to church and worship music in the United States: "Early American Psalmody and the Core Repertory: A Perspective" (D. C. L. Jones) lists the 51 individual core repertory pieces most often printed in America between 1698 and 1819 and discusses significant changes; and "Practicality, Patriotism and Piety: Principal Motivators for Maine Tunebook Compilers, 1794–1830" (L. G. Davenport) discusses characteristics of tunebooks published in Maine around the early nineteenth century. Tables and illustrations; documented with endnotes; index.

179. Pratt, Waldo Selden. *Musical Ministries in the Church: Studies in the History, Theory and Administration of Sacred Music.* 6th ed. New York: G. Schirmer, 1923. 213 p. ML3000 .P96 1923; reprint, New York: AMS Press, 1976. ISBN 0–404–13095-X ML3000 .P91 1976

Five chapters cover religion and the art of music, hymns and hymn singing, the choir, the organ and the organist, the minister's responsibility, and the history of English hymnody. Classified bibliography of about sixty books on church music in general, seventy-five books on hymns and hymn writers, and fifty American church hymnals published since 1880.

180. Rice, William C. *A Concise History of Church Music.* New York: Abingdon Press, 1964. 128 p. ML3000 .R5

Quite concise. Purpose: to "trace the main stream of church music from its earliest beginnings and . . . bring in only those secondary streams that made significant contributions to its growth." Three chapters on music in the United States: Chapter 6, "America, 1600–1800," Chapter 8, "The Century of [Lowell] Mason and [Dwight Lyman] Moody," and Chapter 10, "The Twentieth Century in America." General bibliography with five writings related to American music.

181. Ritter, Frédéric Louis. *Music in America.* 2nd ed. Introduction by Johannes Riedel. New York: C. Scribner's Sons, 1890 (rep. 1895, 1900). xiv, 521 p. ML200 .R62; reprint, New York: Johnson Reprint Corp., 1970. xviii, xiv,

521 p. ML200 .R62 1970; reprint, New York: Burt Franklin, 1972. xiv, 521 p. ISBN 0–8337–3004–5 ML200 .R62 1972

American music, 1620 to 1880. Originally published in 1883, revised in 1890 with several reprints. First five chapters discuss church and worship music. Topics include Puritan psalmody, William Billings and contemporaries, and singing schools. Musical examples and tables; expansive index.

182. Routley, Erik. *Twentieth Century Church Music.* New York: Oxford University Press; London: Jenkins, 1964 (rep. 1966, 1971). 244 p. ISBN 0–19–519162–5 (OUP, 1971). ML3131 .R68; Carol Stream, Ill.: Agape, 1984. ISBN 0–916642–23–2 ML3131 .R68 1984

Survey of twentieth-century Protestant church music. Primarily on English music, but includes a chapter on American music. Musical examples; index of composers and titles; discography of approximately fifty recordings; index.

183. Sablosky, Irving. *American Music.* Chicago: University of Chicago Press, 1969. xiii, 228 p. ISBN 0–226–73324–6 ML200 .S23; reprint, Chicago: University of Chicago Press, 1985. ISBN 0–226–73326–2 ML200 .S23 1985

Historical survey of music in America. Chapter 1 covers Puritan music, *The Bay Psalm Book* (1640), and singing schools in New England. Chapter 3 covers religious folk hymns, shape-note singing, and spirituals. Chapter 4 discusses the contributions of Lowell Mason, John S. Dwight, and contemporaries. The remainder of the book is primarily concerned with secular music. Chronology; annotated bibliography of recommended writings; annotated discography of recordings; expansive index.

184. Schalk, Carl, ed. *Key Words in Church Music: Definition Essays on Concepts, Practices, and Movements of Thought in Church Music.* St. Louis, M.: Concordia, 1978. 365 p. ISBN 0–570–01317–8 ML102.C5 K5

Not a dictionary. Seventy-six essays by thirty contributors. Grouped into fifty-six topics; several essays relate to U.S. church music. Recommended articles are: "Anthem" (E. Wienandt); "Caecilian Movement" (E. Pfeil); "Church Music History: American" (L. Ellinwood); "Church Music History: American Lutheran" (E. Wolf); "Gospel Song" (C. R. Young); "Hymnody: American" (Ellinwood); "Hymnody: American Lutheran" (Schalk); and some mention of U.S. music in articles pertaining to Moravian church music, organ music, the service, and theology. Each essay concludes with suggested readings. Companion to *A Handbook of Church Music* edited by Carl Halter and Schalk (item 565). Cross-references between essays as well as between the two companion books. Graphs, illustrations, and musical examples.

185. Shaw, Benjamin. *Studies in Church Music.* 2nd ed. S.1.: Benjamin Shaw, 1994. 28 p. ML3186 .S534 1994

Church music practices within the following five areas: congregation, choir, instrument, content, and style. Brief discussion of metrical psalmody. Bibliography of twelve writings.

186. Squire, Russel N. *Church Music: Musical and Hymnological Developments in Western Christianity.* St. Louis, Mo.: Bethany Press, 1962. 317 p. ML3000 .S77

Chapter 6, "Religious Music in the United States of America," covers: Spanish and French background; Colonial days; Quaker, Mennonite, Moravian, and other influences; the Church of England; musical personages; late-nineteenth-century artistic religious music; gospel song; the spiritual; and shaped-note singing schools. Chapter 5, "The Organ," offers a little information late in the chapter under the subheading, "The Organ in the New World." Musical examples and illustrations; classified bibliography of 124 writings, with nineteen specifically about music in the United States; index.

187. Stevenson, Robert Murrell. "Church Music: A Century of Contrasts" in *One Hundred Years of Music in America.* Edited by Paul Henry Lang. New York: G. Schirmer, 1961. pp. 80–108. ML200 .L25; reprint, New York: Da Capo Press, 1985. ISBN 0–306–76343–0 ML200 .O53 1985

Examines diversity within church music in the United States between 1861 and 1961. Surveys numerous significant church musicians and discusses briefly their contributions to the musical scene. Documented with endnotes.

188. Stevenson, Robert Murrell. *Protestant Church Music in America: A Short Survey of Men and Movements from 1564 to the Present.* New York: W. W. Norton, 1966 (rep. 1970). xiii, 168 p. ML3111 .S83; ML3111 .S83 1970

Purpose: "to provide a compressed text for use in seminaries, choir schools, and colleges." Topics include: early contacts with the aborigines; New England Puritanism (1620–1720); "regular singing" (1720–1775); Pennsylvania Germans, such as German Baptists (Dunkers), German Pietists, Moravians, and the Ephrata Cloister; native-born composers in the middle Atlantic colonies; the South before 1800; singing-school masters in the New Republic; the half-century preceding the civil war; Negro spirituals, origins and present-day significance; and diverging currents (1850–present). Musical examples and facsimiles; bibliography of approximately 250 items; expansive index.

189. Tawa, Nicholas E. *High-Minded and Low-Down: Music in the Lives of Americans, 1800–1861.* Boston, Mass.: Northeastern University Press,

2000. xiii, 350 p. ISBN 1–55553–443–0, ISBN 1–55553–442–2 (pbk.) ML3917.U6 T39 2000

Examination of American musical culture, sacred and secular, from 1800 to 1861. Church and worship music is included in the following chapters and their subsections: Chapter 3, "Becoming Acquainted with Music" (Discovering Sacred Music outside the Home), Chapter 5, "Amateur Music Making at Home" (Religious Music within the Home Circle), and Chapter 7, "Education and Religion" (Sunday Schools; White Church Services and Prayer Meetings; Black Church Services and Prayer Meetings; and Camp Meetings). General bibliography of about 350 writings; expansive index.

190. Temperley, Nicholas. "Worship Music in English-Speaking North America, 1608–1820" in *Taking a Stand: Essays in Honour of John Beckwith.* Edited by Timothy J. McGee. Toronto: University of Toronto Press, 1995. pp. 166–184. ISBN 0–8020–0583–7 ML205.1 .T35 1995

Details various types of music performed in worship in Canada and the United States. Examines the musical practices of the Episcopal, Congregationalist, Presbyterian, Baptist, Methodist, Dutch Reformed, and Catholic churches. Classified bibliography of nearly forty writings.

191. Westermeyer, Paul. *Te Deum: The Church and Music: A Textbook, a Reference, a History, an Essay.* Minneapolis, Minn.: Fortress Press, 1998. xv, 412 p. ISBN 0–8006–3146–3 BV290 .W47 1998

Not a book about the Te Deum, per se; historical overview of religious music from approximately 1750 B.C. to the 1990s. Three chapters, "American Developments," "Revivalism, Liturgical Renewal, and Spirituals," and "Recurrent American Themes and Richer Textures" specifically address church music in the United States from the American Revolution through the twentieth century. Topics include: singing schools; hymns, spirituals, and African-American music; organs, vocal ensembles, and choirs; various religious groups and music societies; and contributing composers and their music. Chronology; bibliography of nearly seven hundred writings; expansive general index and an index to biblical references.

192. Wilson-Dickson, Andrew. *A Brief History of Christian Music: From Biblical Times to the Present.* Oxford, England: Lion Publishing, 1997. 480 p. ISBN 0–7459–3773-X ML3000 .W54 1997

Revision of the author's *The Story of Christian Music: From Gregorian Chant to Black Gospel: An Authoritative Illustrated Guide to All the Major Traditions of Music and Worship* (Lion Publishing, 1992; Augsburg Fortress, 1996). Surveys the development of Christian music from Gregorian chant to present times. Nine parts: one part (forty-six pages) is dedicated to music in North America from 1600 to the 1990s; further subdivided into following topics: "Christianity Comes to the New World," "Africans in America," "Gospel Music: White and Black," and "The United States and the European Classical Tradition." Introductory in nature; lacks detailed coverage. Some musical examples and facsimiles, but fewer than the musical examples, facsimiles, illustrations, and photos present in its predecessor, *The Story of Christian Music.* General bibliography and discography with approximately twenty writings and sixteen recordings relating to U.S. music; glossary of terms; expansive index.

See also: Gleason, Harold, and Warren Becker. *Early American Music: Music in America from 1620 to 1920* (item 23)

V

Regional Studies

MID-ATLANTIC STATES (DELAWARE, MARYLAND, NEW JERSEY, NEW YORK, PENNSYLVANIA, WASHINGTON, D.C.)

General Works

See: Gillespie, Paul F., ed. *Foxfire 7* (item 220); Maultsby, Portia K. "Music of Northern Independent Black Churches during the Ante-Bellum Period" (item 233)

Maryland

See: Silverberg, Ann Louise. "Cecilian Reform in Baltimore, 1868–1903" (item 293)

New Jersey

See: Robinson, Henrietta Fuller, and Carolyn Cordelia Williams. *Dedicated to Music: The Legacy of African American Church Musicians and Music Teachers in Southern New Jersey, 1915–1990* (item 42)

New York

193. McDaniel, Stanley Robert. "Church Song and the Cultivated Tradition in New England and New York." D.M.A. dissertation. Los Angeles: University of Southern California, 1983. 3 vols.: xiii, 1004 p.

Study of sacred musical practices in nineteenth-century New England and New York churches. Introduced by brief survey of the church song before 1900. Topics include choral music in the church, congregational singing, boy choirs, mixed and chorus choirs, and music ministry and professional church musicians. Several appendixes: brief biographical sketches and lists of representative works of ninety U.S. composers active in region during the nineteenth century. Musical examples, illustrations, and tables; bibliography of nearly seven hundred writings and a few interviews.

See also: Boyer, Horace Clarence. "An Analysis of Black Church Music with Examples Drawn from Services in Rochester, New York" (item 226); Koskoff, Ellen. *Music in Lubavitcher Life* (item 321); Reynolds, William Jensen. *Baptist Hymnody in America: From Roger Williams to Samuel Francis Smith* (item 256)

Pennsylvania

194. Carroll, Lucy E. "Three Centuries of Song: Pennsylvania's Choral Composers, 1681–1981." D.M.A. dissertation. Philadelphia: Combs College of Music, 1982. xvi, 335 p.

Study of sacred and secular choral music in Pennsylvania, 1681 to 1981. Separate chapters on the choral music of the Wissahickon Hermits, Johann Conrad Beissel and the Ephrata Cloister, the Bethlehem Moravians, hymn writers, the choral spiritual, and the Catholic choral revival. Musical examples, facsimiles, and photos; chronology; bibliography of around 180 writings, forty hymnals and choral collections, and 110 scores and recordings.

195. Gaul, Harvey B. "Three Hundred Years of Music in Pennsylvania" in *Music and Musicians of Pennsylvania.* Edited by Gertrude Martin Rohrer. Philadelphia: Theodore Presser, 1940. pp. 46–90. ML200.7.P3 R6; reprint, Port Washington, N.Y.: Kennikat Press, 1970. ISBN 0–87198–512–8 ML200.7.P3 R6 1970

Music in Pennsylvania from 1637 to approximately 1940. Includes discussion of the music of Native Americans, Swedish immigrants, the Quakers, German Pietists, Moravians, Jews, and other religious groups.

196. Grider, Rufus A. *Historical Notes on Music in Bethlehem, Pennsylvania from 1741 to 1871.* Philadelphia: John L. Pile, 1873. 41 p. ML200.8.B56 G8; reprint, Winston-Salem, N.C.: Moravian Music Foundation, 1957. Foreword by Donald M. McCorkle. ML200.8.B56 G8 1957

Interesting view of musical development among Moravians in Bethlehem, Pennsylvania, during the eighteenth and nineteenth centuries by an author who lived during the latter years of that period. Treats sacred and secular music. See comments about music instruction on p. 7, Whit Monday on

p. 8, church organists on p. 10, church music on p. 11, Passion week and Maundy Thursday on p. 14, and Easter on p. 18.

197. National Society of the Colonial Dames of America. *Church Music and Musical Life in Pennsylvania in the Eighteenth Century.* Philadelphia: National Society of the Colonial Dames of America, 1926–1947. 3 vols. in 4: xiii, 261 p.; xii, 291 p.; xiii, 4–217 p.; xiv, 222–576 p. ML3111.P3 N18; reprint, New York: AMS Press, 1972. ISBN 0–404–08090–1 ML3111.P3 N18 1972

Three volumes, third volume published in two parts. Includes some secular topics, but mostly relates to church music. Discusses music of the Episcopal Church, Jewish Synagogues, German Baptists (Dunkers), Lutherans, Mennonites, Moravians, Reformed Church, Catholic church, Schwenkfelders, and other religious groups. All volumes rich in illustrations, photographs, and facsimiles; vol. 1, bibliography of about seventy-five writings; vol. 2, bibliography of about one hundred writings; vol. 3, classified bibliography of about 150 writings; each volume has expansive index.

See also: Fox, Pauline Marie. "Reflections on Moravian Music: A Study of Two Collections of Manuscript Books in Pennsylvania ca. 1800" (item 346); Hall, Harry H. "The Moravian Wind Ensemble: Distinctive Chapter in America's Music" (item 348); Larson, Paul. *An American Musical Dynasty: A Biography of the Wolle Family of Bethlehem, Pennsylvania* (item 39); Martin, Betty Jean. "The Ephrata Cloister and its Music, 1732–1785: The Cultural, Religious, and Bibliographical Background" (item 261); Runner, David Clark. "Music in the Moravian Community of Lititz" (item 354); Sachse, Julius Friedrich. *The Music of the Ephrata Cloister; Also, Conrad Beissel's Treatise on Music as Set Forth in a Preface to the "Turtel Taube" of 1747, Amplified with Fac-Simile Reproductions of Parts of the Text and Some Original Ephrata Music of the Weyrauchs Hügel, 1739; Rosen und Lilien, 1745; Turtel Taube, 1747; Choral Buch, 1754, etc.* (item 262); Westermeyer, Paul. "What Shall We Sing in a Foreign Land?: Theology and Cultic Song in the German Reformed and Lutheran Churches of Pennsylvania, 1830–1900" (item 311)

MIDWESTERN STATES (ILLINOIS, INDIANA, IOWA, KANSAS, MICHIGAN, MINNESOTA, MISSOURI, NEBRASKA, NORTH DAKOTA, OHIO, SOUTH DAKOTA, WISCONSIN)

Illinois

See: Wright, Jeremiah A. "Black Sacred Music: Problems and Possibilities" (item 243)

Michigan

See: Savaglio, Paula Clare. "Polish-American Music in Detroit: Negotiating Ethnic Boundaries" (item 292)

Missouri

198. Krohn, Ernst C. *Missouri Music.* New York: Da Capo Press, 1971. xl, 380 p. ISBN 0–306–70932–5 ML200.7.M47 K72

Collection of essays by the author. Of note are: "*The Missouri Harmony:* A Study in Early American Psalmody" (originally published in 1949), "A Check List of Editions of *The Missouri Harmony*" (originally, 1950), and "Bach Renaissance in St. Louis" (originally, 1955), which discusses performances of Johann Sebastian Bach's music in auditoriums and churches in St. Louis. Facsimiles; documented with endnotes; general index.

Ohio

See: Christenson, Donald Edwin. "Music of the Shakers from Union Village, Ohio: A Repertory Study and Tune Index of the Manuscripts Originating in the 1840's" (item 375); Hartzell, Lawrence W. *Ohio Moravian Music* (item 349); Smith, Harold Vaughn. "Oliver C. Hampton and Other Shaker Teacher-Musicians of Ohio and Kentucky" (item 383); Wetzel, Richard D. *Frontier Musicians on the Connoquenessing, Wabash, and Ohio: A History of the Music and Musicians of George Rapp's Harmony Society (1805–1906)* (item 312)

NEW ENGLAND (CONNECTICUT, MAINE, MASSACHUSETTS, NEW HAMPSHIRE, RHODE ISLAND, VERMONT)

General Works

199. Benes, Peter, and Jane Montague Benes, eds. *New England Music: The Public Sphere, 1600–1900.* Boston, Mass.: Boston University, 1998. 208 p. ML200.7.N52 N48 1998

A collection of fourteen essays by various authors on New England music, seventeenth through nineteenth centuries. Essays relating to church and worship music: "Thomas Walter and the Society for Promoting Regular Singing in the Worship of God: Boston, 1720–1723" (A. C. Buechner); "Singing and Reading: Cooper's Public Presentation of Psalmody in *The*

Last of the Mohicans" (C. C. Boots); "Christmas Religious Music in Eighteenth-Century New England" (S. Nissenbaum); "Evangelical Hymns and Popular Belief" (S. A. Marini); "The *Young Convert's Pocket Companion* and Its Relationship to Migration Patterns of American Religious Folk Song" (E. Laurance); and "The Power of Music Enhanced by the Word: Lowell Mason and the Transformation of Sacred Singing in Lyman Beecher's New England" (M. Dennis Burns). Musical examples and photos; documented with footnotes; general, classified bibliography of writings, ninety-four relating to psalmody, hymnody, and singing schools.

200. Clarke, Garry E. *Essays on American Music.* Westport, Conn.: Greenwood Press, 1977. xviii, 259 p. ISBN 0–8371–9484–9 ML200.1 .C6

Chapter 1, "The Yankee Tunesmiths," discusses the contributions of the "First New England School," namely Asahel Benham, Supply Belcher, William Billings, Jacob French, Oliver Holden, Andrew Law, Justin Morgan, Daniel Read, and Timothy Swan. Documented with endnotes; facsimiles and musical examples; general bibliography of about 260 writings; general index.

201. Tawa, Nicholas E. *From Psalm to Symphony: A History of Music in New England.* Boston, Mass.: Northeastern University Press, 2001. xiv, 466 p. ISBN 1–55553–491–0 ML200.7.N3 T39 2001

Survey of the musical life in New England from *The Bay Psalm Book* (1640) to the end of the twentieth century. The beginning portion of this study broadly addresses musical practice in New England, namely psalms, singing schools, the use of organ, and the contributions of significant individuals. The larger portion of this study, however, is primarily devoted to secular music. Illustrations; general bibliography of nearly 250 writings; expansive index.

202. Thompson, James William. "Music and Musical Activities in New England, 1800–1838." Ph.D. dissertation. Nashville, Tenn.: Peabody College at Vanderbilt University, 1962. xi, 673 p.

Chapter 2 presents background information about sacred music from the early 1700s to 1800. Chapter 3 surveys musical activities in New England from 1800 to 1838, dividing the study into four decades: 1800–1809, 1810–1819, 1820–1829, and 1830–1838. Discusses significant reforms to sacred music, publications of sacred music, the contributions of Lowell Mason (1792–1872), singing schools and church choirs, musical societies and public concerts, and organs, organ building, and organists. Chapters 5 and 6, respectively, treat literary criticism and music education. Discussion of sacred music may be found in these chapters as well. Facsimiles; bibliography of twenty-three published addresses on sacred music; general bibliography of nearly three hundred writings.

See also: Barr, Wayne Anthony. "The History of the Pipe Organ in Black Churches in the United States of America" (item 225); Campbell, Donald Perry. "Puritan Belief and Musical Practices in the Sixteenth, Seventeenth, and Eighteenth Centuries" (item 370); Clark, Linda J. *Music in Churches: Nourishing Your Congregation's Musical Life* (item 302); Daniel, Ralph T. *The Anthem in New England before 1800* (item 388); Eskew, Harry, David W. Music, and Paul A. Richardson. *Singing Baptists: Studies in Baptist Hymnody in America* (item 250); Glover, Raymond F., ed. *The Hymnal 1982 Companion* (item 305); Kroeger, Karl. "The Church-Gallery Orchestra in New England" (item 468); Maultsby, Portia K. "Afro-American Religious Music: 1619–1861" (item 232); Maultsby, Portia K. "Music of Northern Independent Black Churches during the Ante-Bellum Period" (item 233); McDaniel, Stanley Robert. "Church Song and the Cultivated Tradition in New England and New York" (item 193); Ode, James A. *Brass Instruments in Church Services* (item 470); Scholes, Percy Alfred. *The Puritans and Music in England and New England: A Contribution to the Cultural History of Two Nations* (item 371); Stoddard, Robert. *Northern Harmony Online Index* (item 726)

Maine

203. Davenport, Linda Gilbert. *Divine Song on the Northeast Frontier: Maine's Sacred Tunebooks, 1800–1830.* Lanham, Md.: Scarecrow Press, 1996. xi, 428 p. ISBN 0–8108–3025–6 ML3186 .D26 1996

Revision of the author's doctoral dissertation, "Maine's Sacred Tunebooks, 1800–1830: Divine Song on the Northeast Frontier" (University of Colorado, 1991). Surveys music collections and composers/compilers in Maine. Historical background drawn from diaries, church records, and other primary sources. Discusses cultural context, important eighteenth- and nineteenth-century composers, and the contents of the music collections. Musical examples, illustrations, and tables; classified index of more than fifty tunebooks and musical collections and more than three hundred writings; index of tunes by Maine compilers in Maine tunebooks, 1794–1830; and an expansive subject index.

204. Edwards, George Thornton. *Music and Musicians of Maine: Being a History of the Progress of Music in the Territory Which Has Come to Be Known as the State of Maine, from 1604 to 1928.* Portland, Maine: Southworth Press, 1928; reprint, New York: AMS Press, 1970. xxv, 542 p. ISBN 0–404–07231–3 ML200.7.M2 E26 1970

Chronological history. A prologue is followed by seven chapters with numerous subsections. Subsections treating church and worship music are many, especially in chapters that cover music before 1820. Subjects covered include: singing of chants on Island of St. Sauveur, early Catholic missionaries as first music teachers, *The Bay Psalm Book* (1640), singing in the meetinghouses, singing societies, lining out of psalms, fuging tunes, notable psalm books, organs and other instruments, and the contributions of William Billings and Supply Belcher. Chapter 7 is a biographical dictionary of approximately twenty-five hundred musicians. Photographs and illustrations; three indexes: biographical, geographical, and expansive general index.

Massachusetts

205. Broyles, Michael. *"Music of the Highest Class": Elitism and Populism in Antebellum Boston.* New Haven, Conn.: Yale University Press, 1992. ix, 392 p. ISBN 0–300–05495–5 ML200.8.B7 B76 1992

Three chapters, "Boston's Place in the American Musical World," "Sacred-Music Reforms in Colonial and Federal America," and "Lowell Mason: Hymnodic Reformer," are fruitful for the study of the sacred music in Boston, Massachusetts, prior to the Civil War. Musical examples and tables; documented with endnotes; bibliography of around 180 writings; expansive general index.

206. Colonial Society of Massachusetts. *Music in Colonial Massachusetts, 1630–1820: A Conference Held by the Colonial Society of Massachusetts, May 17 and 18, 1973.* Boston, Mass.: The Colonial Society of Massachusetts; distributed by the University Press of Virginia, 1980–1985. 2 vols.: xlv, 404 p.; li, 409–1194 p. F61.C71 vol. 53–54 (ML200.7 .M3)

Lengthy essays by various authors: Volume 1: Music in public places; Volume 2: Music in homes and in churches. Latter volume includes several essays on church music in the United States: "Massachusetts Musicians and the Core Repertory of Early American Psalmody" (R. Crawford) surveys significant musicians in the eighteenth and nineteenth centuries with biographical sketches of sixteen; includes bibliography of core repertory of early American psalmody, providing title, composer, number or printings, location and date of first American printing, and date of first known printing; "The Musical Pursuits of William Price and Thomas Johnston" (S. Hitchings) recounts Price's and Johnston's contributions to music in protestant churches, primarily the Episcopal/Anglican Church; and "Eighteenth-Century Organs and Organ Building in New England" (B. Owen). Facsimiles, illustrations, and photographs; bibliographies at the end of some essays; expansive index.

207. Fisher, William Arms. *Notes on Music in Old Boston.* Boston, Mass.: Oliver Ditson, 1918. xvi, 100, ci–cv p. ML200.8.B7 F4; reprint, New York: AMS Press, 1976. ISBN 0–404–12914–5 ML200.8.B7 F4 1976

Brief account of music in Boston from approximately 1630 to 1910. Early chapters, approximately first forty-nine pages, treat church music. Following pages discuss Hay-Market Theatre and the Oliver Ditson Company, the latter a significant music publisher. Illustrations and photos; expansive index.

See also: Cusic, Don. *The Sound of Light: A History of Gospel Music* (item 415); McCorkle, Donald. "Moravian Music in Salem: A German-American Heritage" (item 353); Summit, Jeffrey A. *The Lord's Song in a Strange Land: Music and Identity in Contemporary Jewish Worship* (item 324)

New Hampshire

208. Pichierri, Louis. *Music in New Hampshire, 1623–1800.* Foreword by Otto Kinkeldey. New York: Columbia University Press, 1960. xv, 297 p. ML200.7.N4 P5

Originated as the author's dissertation by the same title (Syracuse University, 1956). Chapter 3, "Religious Music," surveys *The Bay Psalm Book* (1640), psalmody, choirs, a number of important music publications, and descriptions of church music drawn from writings of the time. Chapters presented on significant individuals in the field of church and sacred music—John Hubbard and Samuel Holyoke—and on *The Village Harmony or Youth's Assistant to Sacred Music,* a tunebook with psalm tunes and anthems suitable for worship, but intended for use in schools and singing societies. Musical examples; bibliography of nearly 150 writings and fifty tunebooks; expansive index.

Rhode Island

See: Reynolds, William Jensen. *Baptist Hymnody in America: From Roger Williams to Samuel Francis Smith* (item 256)

SOUTHERN STATES (ALABAMA, ARKANSAS, FLORIDA, GEORGIA, KENTUCKY, LOUISIANA, MISSISSIPPI, NORTH CAROLINA, SOUTH CAROLINA, TENNESSEE, VIRGINIA, WEST VIRGINIA)

General Works

209. Ellington, Charles Linwood. "The *Sacred Harp* Tradition of the South: Its Origin and Evolution." Ph.D. dissertation. Tallahassee: Florida State University, 1969. vi, 164 p.

Historical study of the *Sacred Harp* in relation to the shape-note tradition of the early nineteenth century through the 1960s. Examines the role of Benjamin Franklin White, one of the compilers of the first edition, in establishing a singing movement and the influence of the singing movement on singing practices in the southern United States. Map and musical examples; bibliography of 160 writings and a handful of pictorial and audio sources.

See also: Crowder, William S. "A Study of Lined Hymnsinging in Selected Black Churches of North and South Carolina" (item 430); Drummond, R. Paul. *A Portion for the Singers: A History of Music among Primitive Baptists since 1800* (item 368); Eskew, Harry, David W. Music, and Paul A. Richardson. *Singing Baptists: Studies in Baptist Hymnody in America* (item 250); Gillespie, Paul F., ed. *Foxfire 7* (item 220); Maultsby, Portia K. "Afro-American Religious Music: 1619–1861" (item 232); McCarroll, Jesse C. "Black Influence on Southern White Protestant Church Music during Slavery" (item 234); Patterson, Beverly Bush. *The Sound of the Dove: Singing in Appalachian Primitive Baptist Churches* (item 369); Stevenson, Arthur Linwood. *The Story of Southern Hymnology* (item 223)

Alabama

210. Willett, Henry, ed. *In the Spirit: Alabama's Sacred Music Traditions.* Montgomery, Ala.: Black Belt Press for the Alabama Folklife Association, 1995. 127 p. ISBN 1–881320–54–5 ML2911.7.A2 I5 1995

A collection of 12 essays by various authors on sacred music in Alabama. Essays: "The African-American Covenanters of Selma, Alabama" (Willett); "The Moan-and-Prayer Event in African-American Worship" (W. Collins); "Singing 'Dr. Watts': A Venerable Hymn Tradition among African Americans in Alabama" (J. Cauthen); "Sand Mountain's Wootten Family: Sacred Harp Singers" (B. Cobb); "Judge Jackson and the Colored Sacred Harp" (Willett); "The Deasons: A Christian Harmony Family" (A. H. F. Kimzey); "Seven-Shape-Note Gospel Music in Northern Alabama: The Case of the Athens Music Company" (C. Wolfe); "Shape-Note Gospel Singing on Sand Mountain" (Cauthen); "Of Related Interest—Convention Gospel Singing in Alabama" (F. C. Fussell); "Community and the Jefferson County, Alabama, Gospel Quartet Tradition" (D. Seroff); "Cry Holy unto the Lord: Tradition and Diversity in Bluegrass Gospel Music" (J. Bernhardt); and "Of Related Interest—Margie Sullivan: Mother of Bluegrass Gospel" (E. Kellen). Photographs and facsimiles. Accompanied by a compact disc.

Georgia

211. Byrnside, Ronald L. *Music in Eighteenth-Century Georgia.* Athens: University of Georgia Press, 1997. xiii, 146 p. ISBN 0–8203–1853–1 ML200.7.G4 B89 1997

Three chapters concentrate on church and worship music: Chapter 4, "Religious Groups in Georgia," discusses the music of the Episcopal church, Salzburgers and Moravians, Italians, French, and Spanish Jews, and other protestant groups; Chapter 5, "John Wesley in Georgia," discusses Wesley's brief career in Georgia, covering metrical psalmody and the German chorale; and Chapter 6, "George Whitefield and the Great Awakening," concerns Whitefield's visits to Georgia, primarily Savannah and Charleston. Other topics on church and worship music can be accessed through the expansive index. Musical examples; bibliography of about 135 writings.

212. Hoogerwerf, Frank W., ed. *Music in Georgia.* New York: Da Capo Press, 1984. xvi, 343 p. ISBN 0–306–76096–7 ML200.7.G4 M9 1984

A collection of twenty-four essays by various authors previously published in journals. Essays on church and worship music: "The Sacred Harp in the Land of Eden" (D. Davidson); "Lowell Mason's Varied Activities in Savannah" (M. Freeman LaFar); "Wanted: An American Hans Sachs" (G. P. Jackson) describes the Southeastern area of the United States as a significant center of music, with Georgia and South Carolina as its earliest development; "John Wesley's First Hymnbook" (R. Stevenson) discusses his *Collection of Psalms and Hymns* (1737) and following editions; "Salem Camp Meeting: Symbol of an Era" (W. B. Garrison); "Early Sounds of Moravian Brass Music in America: A Cultural Note from Colonial Georgia" (H. H. Hall); and "The Royal Singing Convention, 1893–1931: Shape Note Singing Tradition in Irwin County, Georgia" (K. L. Jackson). Some essays documented with endnotes; index.

See also: Hall, Harry H. "The Moravian Wind Ensemble: Distinctive Chapter in America's Music" (item 348); Lawson, Charles T. "Musical Life in the Unitas Fratrum Mission at Springplace, Georgia, 1800–1836" (item 356); Norton, Kay. *Baptist Offspring, Southern Midwife: Jesse Mercer's* Cluster of Spiritual Songs *(1810): A Study in American Hymnody* (item 465); Sacré, Robert, ed. *Saints and Sinners: Religion, Blues, and (D)evil in African-American Music and Literature: Proceedings of the Conference Held at the Université de Liège (October 1991)* (item 237)

Kentucky

213. Carden, Joy. *Music in Lexington before 1840.* Preface by James A. Ramage. Lexington, Ky.: Lexington-Fayette County Historical Commission, 1980. xi, 148 p. ML200.8.L5 C4

Music in Lexington, Kentucky, late 1700s to 1840. Chapter 6, "Church Music: Hymn Books, Revivals, and Camp Meetings," although rather brief, covers topics in the chapter's subtitle along with organ and funeral music.

Chapter 4, "Music Education: Singing Schools, Private Lessons and Music Academies," discusses psalmody and shape note singing. Music facsimiles, illustrations, and tables; documented with endnotes; index.

214. Montell, William Lynwood. *Singing the Glory Down: Amateur Gospel Music in South Central Kentucky, 1900–1990.* Lexington: University Press of Kentucky, 1991. xi, 248 p. ISBN 0–8131–1757–7 ML3187 .M66 1991

Montell, William Lynwood. *Singing the Glory Down: Amateur Gospel Music in South Central Kentucky, 1900–1990.* Owensboro, Ky.: Owensboro Volunteer Recording Unit, 1991. 1 audio cassette.

Examines "the rich history and performance aspects of community-based shape-note gospel music and its offshoots—church singings, singing conventions, and four-part southern harmony quartet music" in the twentieth century. Includes a list of the region's singing groups and a list of shape-note teachers in the area. Photographs and map; bibliography of approximately eighty interviews and fifty writings; expansive index. Also published on cassette.

See also: Leist, Stephen G. *The History of Music at Christ Church Cathedral: 1796–1996* (item 306); Smith, Harold Vaughn. "Oliver C. Hampton and Other Shaker Teacher-Musicians of Ohio and Kentucky" (item 383)

Mississippi

See: Hitt, Gwen Keys. *We Shall Come Rejoicing: A History of Baptist Church Music in Mississippi* (item 252)

North Carolina

See: Hall, Harry H. "The Moravian Wind Ensemble: Distinctive Chapter in America's Music" (item 348)

South Carolina

See: Sacré, Robert, ed. *Saints and Sinners: Religion, Blues, and (D)evil in African-American Music and Literature: Proceedings of the Conference Held at the Université de Liège (October 1991)* (item 237)

Virginia

215. Stoutamire, Albert. *Music of the Old South: Colony to Confederacy.* Rutherford, Va.: Fairleigh Dickinson University Press, 1972. 349 p. ISBN 0–8386–7910–2 ML200.7.V5 S8

Expansion of the author's doctoral dissertation, "A History of Music in Richmond, Virginia from 1742 to 1855" (Florida State University, 1960). Musical activities in Virginia from the late 1600s to 1865. Chapters 2 through 5 cover the years 1780–1799, 1800–1825, 1826–1845, and 1846–1865, respectively. Within each of these chapters, there is a subsection on church music. Illustrations and photos; bibliography of around 125 writings; expansive index.

See also: Eskew, Harry. "Shape-Note Hymnody in the Shenandoah Valley, 1816–1860" (item 463); Titon, Jeff Todd. *Powerhouse for God: Speech, Chant, and Song in an Appalachian Baptist Church* (item 258)

SOUTHWESTERN STATES (ARIZONA, NEW MEXICO, OKLAHOMA, TEXAS)

General Works

216. Swan, Howard. *Music in the Southwest: 1825–1950.* Preface by Robert Glass Cleland. San Marino, Calif.: Huntington Library, 1952. x, 316 p. ML200.7.S74 S9; reprint, New York: Da Capo Press, 1977. ISBN 0–306–77418–6 ML200.7.S74 S9 1977

Examines music of southwestern and western United States, spanning California, Arizona, Nevada, and Utah. Early chapters discuss worship music of the Mormon church and the music of the mission, rancho, and pueblo of California. Music of Native Americans and early Spanish period in California are omitted. Later chapters examine secular music. Photos and illustrations; bibliography of nearly 170 writings; expansive index.

See also: Drummond, R. Paul. *A Portion for the Singers: A History of Music among Primitive Baptists since 1800* (item 368)

Oklahoma

217. Adams, K. Gary. "Music in the Oklahoma Territory: 1889–1907." Ph.D. dissertation. Denton: University of North Texas, 1979. x, 226 p.

One chapter (twenty-four pages) devoted to service music, organ recitals, and concerts in churches within the Oklahoma Territory around the turn of the twentieth century. General bibliography of forty-three writings.

Texas

See: Lester, Joan Stadelman. "Music in Cumberland Presbyterian Churches in East Texas Presbytery, 1900–1977, as Recorded in Church

Records and as Related in Oral and Written Interviews" (item 367); *Sacred Harp Singing in Texas* (item 725)

WESTERN STATES (ALASKA, CALIFORNIA, COLORADO, HAWAII, IDAHO, MONTANA, NEVADA, OREGON, UTAH, WASHINGTON, WYOMING)

California

218. Koegel, John. "Spanish Mission Music from California: Past, Present and Future Research." *The American Music Research Center Journal,* 3 (1993): 78–111

Sacred and secular Spanish music in California prior to American annexation in 1846. Reviews research of Sister Mary Dominic Ray culminated in her *Gloria Dei: The Story of California Mission Music* (1974) (item 291), along with recent discoveries, and offers recommendations for future research. Musical examples, facsimiles, and photos; classified bibliography of nearly 230 writings (mission music in California and New Mexico; Hispanic-American (non-mission) Catholic sacred music; Hispanic-American and Mexican religious musical folk theater; Hispanic-American, Mexican, and Indian folk music and folklore; and social and cultural life of Hispanic southwest).

See also: McGann, Mary E. "Interpreting the Ritual Role of Music in Christian Liturgical Practice" (item 286); Ray, Mary Dominic, and Joseph H. Engbeck. *Gloria Dei: The Story of California Mission Music* (item 291)

Hawaii

219. Stoneburner, Bryan C. *Hawaiian Music: An Annotated Bibliography.* New York: Greenwood Press, 1986. x, 100 p. ISBN 0–313–25340–4 ML125.H4 S76 1986

An annotated bibliography of 564 resources on Hawaiian music, covering 1831 through 1985. The index proves an invaluable tool for locating resources on chant, Hawaiian hymns, religious aspects of music, and so on. Glossary of Hawaiian terms.

VI

Religious and Ethnic Groups

GENERAL WORKS

220. Gillespie, Paul F., ed. *Foxfire 7.* Garden City, N.Y.: Anchor Press/Doubleday, 1982. 510 p. ISBN 0–385–15243–4, ISBN 0–385–15244–2 (pbk.) BR563.M68 F69 1982

    Gillespie, Paul F., ed. *Foxfire 7.* Lexington, Ky.: Lexington Volunteer Recording Unit, 2003. 14 audio cassettes.

    Collection of essays and personal accounts of religion in rural, southern Appalachia. Several denominations are represented: Baptist, Catholic, Church of Christ, Episcopal, Methodist, Pentecostal, and Presbyterian. Essays/personal accounts related to music include: "The Tradition of Shaped-Note Music: A History of Its Development" (E. Card); "Christian Harmony" (R. Moss, Q. Smathers); "'I Love to Sing . . . '" (various); and "Gospel Shaped Note Music" (J. Raby, H. Fountain, A. Norton, L. Smith, H. Page, H. Brown, E. Pitts, E. Watts, F. Watts, W. Maney). Facsimiles and photos; glossary of terms; documented with endnotes; expansive index. Published as a cassette tape audio book in 2003.

221. Liemohn, Edwin. *The Organ and Choir in Protestant Worship.* Philadelphia: Fortress Press, 1968. x, 178 p. ML3100 .L54

    History of the use of choirs and organs in Protestant churches from pre-Reformation through present. Emphasis upon Lutheran, Episcopal/Anglican, and Presbyterian music with frequent mention of Baptist, Methodist, and Moravian music. Concentrates geographically upon Scandinavian countries,

Germany, Switzerland, the Netherlands, the British Isles, and the United States Bibliography of 217 items; expansive index.

222. Lornell, Kip, and Anne K. Rasmussen, eds. *Musics of Multicultural America: A Study of Twelve Musical Communities.* New York: Schirmer Books; London: Prentice Hall International, 1997. xii, 348 p. ISBN 0–02–864585–5 ML3477.M88 1997

Thirteen essays, mostly on secular topics, by various authors. Essays relating to sacred music in the United States are: "Cultural Interaction in New Mexico as Illustrated in the Metachines Dance" (B. M. Romero), which examines music and dance of the native Pueblo and Mexican American citizens of New Mexico and southern Colorado; "*Waila:* The Social Dance Music of the Tohono O'odham" (J. S. Griffith) which studies performance practice of post-World War II sacred/secular *waila* music in southern Arizona by the Tohono O'odham, also known as the Pima Indians; "Triangles, Squares, Circles, and Diamonds: The 'Fasola Folk' and Their Singing Tradition" (R. Pen) which traces the development of psalm singing, shaped-note singing, white spirituals, and camp meeting hymnody; and "The Memphis African American Sacred Quartet Community" (Lornell) which explores the relationship between the religious and popular aspects of the music. Almost all essays include photographs, bibliographies, and discographies; some include glossaries; general, expansive index.

223. Stevenson, Arthur Linwood. *The Story of Southern Hymnology.* Salem, Va.: Arthur L. Stevenson; Roanoke, Va.: Printed by the Stone Printing and Manufacturing Co., 1931. vi, 187 p. ML3111.S8; reprint, New York: AMS Press, 1975. ISBN 0–404–08334-X ML3111.S8 1975

Originally published in 1931. Southern hymnody from the eighteenth century to the early decades of the twentieth century. Addresses musical practices of Baptist, Presbyterian, and Methodist churches. Discusses gospel hymns, early Sunday school hymns, singing schools, among other topics. Tables; no index.

See also: Byrnside, Ronald L. *Music in Eighteenth-Century Georgia* (item 211); Christenson, Donald Edwin. "Music of the Shakers from Union Village, Ohio: A Repertory Study and Tune Index of the Manuscripts Originating in the 1840's" (item 375); Gaul, Harvey B. "Three Hundred Years of Music in Pennsylvania" in *Music and Musicians of Pennsylvania* (item 195); National Society of the Colonial Dames of America. *Church Music and Musical Life in Pennsylvania in the Eighteenth Century* (item 197)

AFRICAN AMERICAN

224. Abbington, James, ed. *Readings in African American Church Music and Worship.* Chicago: G. I. A. Publications, 2001. xxi, 594 p. ISBN 1–57999–163–7 ML2911.9 .R42 2001

A collection of forty essays on African-American church music and worship grouped under seven headings: (1) historical perspectives, (2) surveys of hymnals and hymnody, (3) liturgical hymnody, (4) worship, (5) composers, (6) the organ, and (7) contemporary perspectives. Selected essays relating to church music: "Of the Faith of the Fathers" (W. E. B. DuBois); "The Negro Spiritual" (J. W. Work, III); "The Performed Word: Music and the Black Church" (C. E. Lincoln, L. H. Mamiya); "The Use and Performance of Hymnody, Spirituals, and Gospels in the Black Church" (P. K. Maultsby); "African American Song in the Nineteenth Century: A Neglected Source" (I. V. Jackson-Brown); "Hymnals of the Black Church" (E. Southern); "Published Hymnals in the African American Tradition" (M. W. Costen), an annotated bibliography of twenty-nine hymnals published between 1801 and 1987; "Black Hymnody" (W. P. Whalum); "Hymnody of the African American Church" (D. L. White); "The Liturgy of the Roman Rite and African American Worship" (J-Glenn Murray); "The Gift of African American Sacred Song" (T. Bowman); "Music among Blacks in the Episcopal Church: Some Preliminary Considerations" (I. V. Jackson-Brown); "Introduction: Why an African American Hymnal?" (H. T. Lewis); "Hymns and Songs: Performance Notes" (H. C. Boyer); "Service Music: Performance Notes" (C. Haywood); "Worship and Culture: An African American Lutheran Perspective" (J. A. Donella, II, J. Nunes, K. M. Ward); "Leading African American Song" (M. V. Burnim); "Worship Activities" (B. E. Mays, J. W. Nicholson); "Some Aspects of Black Worship" (C. G. Adams); "The Liturgy of Zion: The Soul of Black Worship" (W. B. McClain); "Definitions of Praising and a Look at Black Worship" (B. E. Aghahowa); "Indicted" (V. M. McKay); "Church Music by Black Composers: A Bibliography of Choral Music" (W. B. Garcia), a bibliography of nearly two hundred choral works, ten writings recommended for the church library, and a list of publishers of choral music; "'Introduction' from *Choral Music by African American Composers: A Selected, Annotated Bibliography*" (E. Davidson White); "Black Composers and Religious Music" (G. Southall); "A History of the Pipe Organ in the Black Church" (W. A. Barr); "The Church Organist, African American Organ Music, and the Worship Service: A Useful Guide" (M. Thomas Terry); "Service Playing for Organists" (compiled and edited by Abbington), a compilation of quotes by musicians on the subject; "What Lies Ahead?" (W. T. Walker); "Conflict and Controversy in Black

Religious Music" (Burnim); "Church Music: A Position Paper (With Special Consideration of Music in the Black Church)" (Whalum); "Music and Worship in the Black Church" (J. W. Mapson, Jr.); "Black Sacred Music: Problems and Possibilities" (J. A. Wright, Jr.); "Christ Against Culture: Anticulturalism in the Gospel of Gospel" (J. M. Spencer); and "'I am the Holy Dope Dealer': The Problem with Gospel Music Today" (O. M. Hendricks, Jr.). Musical examples and bibliographies accompany some essays; no index.

225. Barr, Wayne Anthony. "The History of the Pipe Organ in Black Churches in the United States of America." D.M.A. dissertation. Ann Arbor: University of Michigan, 1999. v, 66 p.

Study of the introduction of the pipe organ into African-American churches with specific emphasis on the history of the pipe organ at St. Thomas' Episcopal Church, Philadelphia, and St. Philip's Episcopal Church, New York. Several appendixes, including a list of organ builders and organs and specifications of selected organs. Bibliography of twenty-eight writings.

226. Boyer, Horace Clarence. "An Analysis of Black Church Music with Examples Drawn from Services in Rochester, New York." Ph.D. dissertation. New York: University of Rochester, 1973. ix, 252 p.

Study of musical practices in African-American churches, including the following groups: Presbyterian, Jehovah's Witnesses, Seventh-Day Adventist, National Spiritualist Association, Church of Christ, Church of God and Saints in Christ, Nation of Islam, United Methodist, African Methodist Episcopal, African Methodist Episcopal Zion, Christian Methodist Episcopal, Baptist, Church of God in Christ, and Church of God. Churches within Rochester, New York, were examined as representative denominational services. Musical examples; bibliography of twenty-three writings and seventeen hymnals.

227. Costen, Melva Wilson, and Darius L. Swann, eds. *The Black Christian Worship Experience.* Rev. and enlarged. Atlanta, Ga.: ITC Press, 1992. 265 p. BR82.7 .B45 1992

A collection of nineteen writings about African-American worship. Two essays relate to music: "Hymnals of The Black Church" (E. Southern) and "The Use and Performance of Hymnody, Spirituals and Gospels in The Black Church" (P. K. Maultsby). Essays documented with footnotes; expansive index.

228. Duncan, Curtis Daniel. "A Historical Survey of the Development of the Black Baptist Church in the United States and a Study of Performance Practices Associated with Dr. Watts Hymn Singing: A Source Book for

Teachers." Ed.D. dissertation. St. Louis, Mo.: Washington University, 1979. viii, 278 p.

Surveys the development of the African-American Baptist church in the United States from its inception in slavery through the antebellum period. Special attention given to the development of Isaac Watts hymn singing, the practice of reciting the hymn text before singing each stanza. Musical examples and facsimiles; bibliography of approximately 135 writings.

229. DuPree, Sherry Sherrod. *African-American Holiness Pentecostal Movement: An Annotated Bibliography.* Preface by Samuel S. Hill. New York: Garland, 1996. lxvii, 650 p. ISBN 0–8240–1449–9 BX8762.5 .D87 1996

More than three thousand writings about the African-American Pentecostal movement in the United States from the 1880s to the mid-1990s. Only brief sections relate to music. In Chapter 1, "Selected Bibliographies on African-American Religion and Culture," three sections treat music dissertations, gospel records, and music books, respectively; Chapter 4, "Apostolic Pentecostal Groups" lists a couple of music books. Helpful appendixes include a glossary of terms, list of denominations, and addresses of organizations and research institutions. General index and geographical index.

230. Epstein, Dena J. *Sinful Tunes and Spirituals: Black Folk Music to the Civil War.* Urbana: University of Illinois Press, 1977 (rep. 1981). xix, 433 p. ISBN 0–252–00520–1, ISBN 0–252–00875–8 (pbk.) ML3556 .E8, ML3556 .E8 1981

African-American folk music in the United States, approximately 1640 to 1867. Three parts: (1) "Development of Black Folk Music to 1800," (2) "Secular and Sacred Black Folk Music, 1800–1867," and (3) "The Emergence of Black Folk Music during the Civil War." Discussion of sacred and secular music. Overview of sacred black folk music, especially during the first half of the nineteenth century. Examines performance practice, the use of instruments, and various types of music and dance against the backdrop of pervading social issues. Musical examples, facsimiles, illustrations, and photographs; bibliography of more than seven hundred writings; expansive index.

231. Jackson, Irene V., ed. *More Than Dancing: Essays on Afro-American Music and Musicians.* Westport, Conn.: Greenwood Press, 1985. xiii, 281 p. ISBN 0–313–24554–1 ML3556 .M68 1985

Eleven essays on African-American music and musicians. Only a few treat the subject of church music or gospel music. Of note: "Music Among Blacks in the Episcopal Church: Some Preliminary Considerations" (Jackson) traces musical activities of African Americans in the Episcopal Church from the late eighteenth century to present day; "A Comparative Analysis of Traditional and Contemporary Gospel Music" (H. C. Boyer) examines lyrics, rhythm and meter, form, accompaniment, vocal timbre and background groups, harmony, and congregational response of "traditional" gospel music—represented by James Cleveland, Shirley Caesar, and Mattie Moss Clark—and "contemporary" gospel music—represented by Andrae Crouch, Danniebelle Hall, and Rahni Harris; and "The Black Gospel Music Tradition: A Complex of Ideology, Aesthetic, and Behavior" (M. V. Burnim), which describes African-American gospel music tradition not as a mere genre but as "a very complex cultural system." Musical examples and charts; bibliographies following essays; expansive general index.

232. Maultsby, Portia K. "Afro-American Religious Music: 1619–1861." Ph.D. dissertation. Madison: University of Wisconsin, 1974. x, 448 p.

Part I: African origins of North American slaves, New England Colonies from the seventeenth through nineteenth centuries, Southern Colonies during the same centuries, religious music and dance in the south during the eighteenth and nineteenth centuries, origin and development of the spiritual and the shout. Part II: Analyses of one hundred African-American spirituals, examining formal structure, textual and melodic structures of first verse and chorus, scale pattern and structure, pitch functions, range, intervallic structure of melodic patterns, three-note and four-note melodic patterns, meter, and rhythmic patterns. Musical examples, facsimiles, and tables; bibliography of about two hundred writings.

233. Maultsby, Portia K. "Music of Northern Independent Black Churches during the Ante-Bellum Period." *Ethnomusicology,* 19/3 (Sept. 1975): 401–420

Examines the development of two religious musical traditions among northern African Americans in the United States prior to the Civil War within the context of social, economical, and political conditions. Bibliography of thirty-seven writings.

234. McCarroll, Jesse C. "Black Influence on Southern White Protestant Church Music during Slavery." Ed.D. dissertation. New York: Columbia University, 1972. iv, 168 p.

Contends that slaves influenced Southern White Protestant church music by means of "(1) travel from one place to another, (2) informal black-white proximity during slavery, [and] (3) organized social and religious interaction between slaves and whites." Facsimiles; bibliography of about three hundred writings.

235. Powell, Deborah Michelle. "A Message of Hope: The Role of Music in the African American Church." M.Div. thesis. Johnson City, Tenn.: Emmanuel School of Religion, 1998. iii, 128 p.

Outlines the role of music in the African-American church from its roots in Africa to the U.S. Civil War. Primarily focused on theological, cultural, and social issues. Discusses gospel music and spirituals but not in great depth. Appendix of approximately fifty-three pages of music; bibliography of forty-seven writings.

236. Riedel, Johannes. *Soul Music, Black & White: The Influence of Black Music on the Churches.* Minneapolis, Minn.: Augsburg, 1975. 159 p. ISBN 0–8066–1414–5 ML3537 .R53

Study of black soul music from its origins in Africa, and white soul music from its origins in Europe. Discusses migration of these influences to North and South America and their impact on music in the United States. Musical examples; bibliography of fifteen writings; discography of thirty recordings.

237. Sacré, Robert, ed. *Saints and Sinners: Religion, Blues, and (D)evil in African-American Music and Literature: Proceedings of the Conference Held at the Université de Liège (October 1991).* Liège, Belgium: Société liégeoise de musicologie, 1996. xii, 352 p. ML3556 .S255 1996

Sixteen essays. Essays relating to religious music: "The Saints and the Sinners under the Swing of the Cross" (Sacré); "The Emergence of the Slave Spirituals in the Georgia-South Carolina Low Country [1830–1870]" (R. M. Lewis); "The Prevailing Spirit of African-American Solo Vocal Music" (M. Thompson); "The Cultural Evolution of the African American Sacred Quartet" (J. M. Jackson); "Barrelhouse Singers and Sanctified Preachers" (C. Lornell); "Denomination Blues: Texas Gospel with Novelty Accompaniment by Washington Phillips" (G. van Rijn); "Prodigal Sons: Son House and Robert Wilkins" (M. Humphrey); "'A Dead Cat on the Line'" (P. Oliver); "The Titanic: A Case Study of Religious and Secular Attitudes in African American Song" (C. Smith); "The Use of Sacred and Secular Music in Rudolph Fisher's 'The Promised Land'" (S. C. Tracy); "Echoes of the Relation between Sacred and Secular Music in the Works of James Baldwin" (P. Anthonissen); "Spiritual and Gospel Music and the Rise of an Afro-American Identity in U.S. Radio" (S. Danchin); "Religious Words in Blues Lyrics and Titles: A Study" (A. J. M. Prévos); and "God's Music vs. the Devil's Music: The Evidence from Blues Lyrics" (R. Springer). Illustrations and tables; general index.

238. Southern, Eileen. *The Music of Black Americans: A History.* 3rd ed. New York: W. W. Norton, 1997. xxii, 678 p. ISBN 0–393–03843–2, ISBN 0–393–97141–4 (pbk.) ML3556 .S74 1997

Musical activities, sacred and secular, of African Americans from 1619 to 1996. Addresses psalmody, hymnody, camp-meeting songs, the spiritual, and the gospel, as well as African-American churches of various denominations. Musical examples, facsimiles, photographs, and illustrations; classified bibliography (general reference; African legacy; eighteenth and early nineteenth centuries; nineteenth century; nineteenth-century collections of music scores; twentieth century; twentieth-century collections of music scores; periodicals; biographies) of more than seven hundred writings and thirty scores; discography (mostly secular music) of nearly forty recordings and sets.

239. Spencer, Jon Michael. *Black Hymnody: A Hymnological History of the African-American Church.* Knoxville: University of Tennessee Press, 1992. xiii, 242 p. ISBN 0–87049–745–6, ISBN 0–87049–760-X (pbk.) BV313 .S64 1992

Hymnological traditions of ten African-American churches in the United States, namely, the African Methodist Episcopal, African Methodist Episcopal Zion, Christian Methodist Episcopal, United Methodist, National Baptists, Church of Christ (Holiness), USA, House of God, Church of God in Christ, Episcopal, and the Catholic churches. Bibliography of twenty-one writings and thirty-five hymnals; music index and expansive general index.

240. Spencer, Jon Michael, ed. *The R. Nathaniel Dett Reader: Essays on Black Sacred Music.* Foreword by Dominique-René de Lerma. Preface and Introduction by Spencer. Durham, N.C.: Duke University Press, 1991. xiv, 138 p. ML3556 .R56 1991

Special issue of *Black Sacred Music: A Journal of Theomusicology,* 5/2 (Fall 1991). Chronologically-arranged writings of Robert Nathaniel Dett (1882–1943), an authority on the African-American spiritual. Dett's writings: "Helping to Lay Foundation for Negro Music of the Future," "The Emancipation of Negro Music," "Development of Negro Secular Music," "The Development of Negro Religious Music," "Negro Music of the Present," "Review of *The Book of American Negro Spirituals,*" "Review of *Negro Workaday Songs,*" "John W. Work," "Religious Folk-Songs of the Negro," "As the Negro School Sings," "Notes to the Hampton Choir at Carnegie Hall," "Notes to the Hampton Choir at Symphony Hall," "Notes to the Hampton Choir at Queen's Hall," "A Musical Invasion of Europe," "From Bell Stand to Throne Room," "The Dett Collection of Negro Spirituals," "Understanding the Negro Spiritual," "The Authenticity of the Spiritual," "The Development of the Negro Spiritual," and "Negro Music." Photos of Dett; bibliography of Dett's published writings; expansive index.

241. Spencer, Jon Michael. *Sing a New Song: Liberating Black Hymnody.* Minneapolis, Minn.: Fortress Press, 1995. viii, 231 p. ISBN 0–8006–2722–9 BV313 .S645 1995

 "Black hymnody" commonly refers to all religious songs within the worship services of African-American churches. Author expresses concern "about hymnic texts that are Eurocentric—exacerbated by European worldview and ideology—to the degree that these texts incur the problems of sexism, racism, and classism." African-American churches are challenged to "sing a new song," breaking away from Eurocentric ideology and creating new musical works with Afro-Christian identity. Musical examples; bibliography of nearly one hundred writings; expansive indexes for names and subjects and for biblical references.

242. Walker, Wyatt Tee. *"Somebody's Calling My Name": Black Sacred Music and Social Change.* Foreword by Gardner C. Taylor. Valley Forge, Pa.: Judson Press, 1979. 208 p. ISBN 0–8170–0849–7 ML3556 .W23

 Covers African roots, oral tradition in the United States, the spiritual, African-American meter music, hymns of improvisation, and gospel music. Musical examples, illustrations, graphs, and tables; documented with endnotes; index.

243. Wright, Jeremiah A. "Black Sacred Music: Problems and Possibilities." D.Min. dissertation. Dayton, Ohio: United Theological Seminary, 1990. 101 p.

 Personalized view of the music of Trinity United Church of Christ, Chicago, Illinois, where music is characteristically African-American sacred music. Addresses spirituals and gospel music. Some historical information regarding African-American churches in the United States Bibliography of thirty-four writings.

See also: Allen, Ray. *Singing in the Spirit: African-American Sacred Quartets in New York City* (item 410); Boyer, Horace Clarence. *How Sweet the Sound: The Golden Age of Gospel* (item 411); Broughton, Viv. *Black Gospel: An Illustrated History of the Gospel Sound* (item 412); Brown, Marian Tally. "A Resource Manual on the Music of the Southern Fundamentalist Black Church" (item 384); Burnim, Mellonee V. "The Black Gospel Music Tradition: Symbol of Ethnicity" (item 413); Burnim, Mellonee V. "The Performance of Black Gospel Music as Transformation" (item 414); Clency, Cleveland Charles. "European Classical Influences in Modern Choral Settings of the African-American Spiritual: A Doctoral Essay" (item 490); Cone, James H. *The Spirituals and the Blues: An Interpretation* (item 491); Crowder, William S. "A Study of Lined Hymnsinging in Selected Black Churches of North and South Carolina" (item 430); Dargan, William T.

"Congregational Gospel Songs in a Black Holiness Church: A Musical and Textual Analysis" (item 314); Dixon, Christa K. *Negro Spirituals: From Bible to Folk Song* (item 492); DjeDje, Jacqueline Cogdell. "A Historical Overview of Black Gospel Music in Los Angeles" (item 416); Jackson, Irene V. "Afro-American Gospel Music and Its Social Setting: With Special Attention to Roberta Martin" (item 419); Lovell, John, Jr. *Black Song: The Forge and the Flame: The Story of How the Afro-American Spiritual Was Hammered Out* (item 493); McGann, Mary E. "Interpreting the Ritual Role of Music in Christian Liturgical Practice" (item 286); Neely, Thomasina. "Belief, Ritual, and Performance in a Black Pentecostal Church: The Musical Heritage of the Church of God in Christ" (item 301); Reagon, Bernice Johnson, ed. *We'll Understand It Better By and By: Pioneering African American Gospel Composers* (item 420); Robinson, Henrietta Fuller, and Carolyn Cordelia Williams. *Dedicated to Music: The Legacy of African American Church Musicians and Music Teachers in Southern New Jersey, 1915–1990* (item 42); Rogal, Samuel J. *A General Introduction to Hymnody and Congregational Song* (item 443); Williams-Jones, Pearl. "Afro-American Gospel Music: A Crystallization of the Black Aesthetic" (item 423)

AFRICAN METHODIST EPISCOPAL

244. Campbell, James T. *Songs of Zion: The African Methodist Episcopal Church in the United States and South Africa.* New York: Oxford University Press, 1995. xv, 418 p. ISBN 0–19–507892–6 BX8443 .C36 1995; reprint, Chapel Hill: University of North Carolina Press, 1998. ISBN 0–8078–4711–9 BX8443 .C36 1998

    Primarily a historical study of the African-American Methodist Episcopal church in the United States and its commitment to missionary work in South Africa. Originated as the author's doctoral dissertation, "Our Fathers, Our Children: The African Methodist Episcopal Church in the United States and South Africa" (Florida State University, 1970). Photographs and illustrations; documented with endnotes; expansive general index. The index heading, "Music, African American," leads to passages about musical practices.

245. Peasant, Julian S. "The Arts of the African Methodist Episcopal Church as Viewed in the Architecture, Music and Liturgy of the Nineteenth Century." Ph.D. dissertation. Athens: Ohio University, 1992. vi, 252 p.

    One chapter (seventy typescript pages) is devoted to music of the African Methodist Episcopal Church. Analyzes fourteen hymns and spiritual songs. Gives brief biographical information about composer and librettist, lists key, meter, and number of measures, provides descriptive remarks about melody, harmony, and rhythm, and discusses the text. Other chapters deal with architecture and liturgy. Musical examples; glossary of terms; bibliography of nearly 130 writings.

246. Southern, Eileen. "The Music of the Black Church in the United States." In *Images de l'Africain de l'Antiquité au XXe Siècle* [Images of the African from Antiquity to the 20th Century], ed. Daniel Droixhe and Klaus H. Kiefer. Frankfurt, Germany: Verlag Peter Lang, 1987. pp. 143–148. ISBN 3–8204–0127-X CB245 .I43 1987

Article in English. Musical practices of the African Methodist Episcopal Church, the first independent African-American denomination, established in 1816. Briefly surveys the use of European and newly composed psalms and hymns, camp meeting hymns, folk music, spirituals, and the gospel song.

AFRICAN METHODIST EPISCOPAL ZION

247. Jones, Marion B. "Music in Liturgy: Enhancing the Quality of Worship in the A. M. E. Zion Church." D.Min. dissertation. Madison, N.J.: Drew University, 1991. ii, 51, 23 p.

Examines the role of music in worship in the African Methodist Episcopal Zion Church. Chapter 3 presents a brief overview of the hymnology movement in the United States. Classified bibliography of more than thirty writings.

ASSEMBLIES OF GOD

248. Tanner, Donald Ray. "An Analysis of Assemblies of God Hymnology." Ph.D. dissertation. Minneapolis: University of Minnesota, 1974. viii, 250 p.

Discusses antecedents and development of music within the Assemblies of God, analyses hymn-tunes in comparison to Baptist and Lutheran church music, and compares the Assemblies of God 1969 Hymnal to the 1930 Hymnal to determine thematic changes in the content. Charts and tables; bibliography of nearly ninety writings and interviews; topical index of hymns.

BAPTIST AND SOUTHERN BAPTIST

249. *Church Music in Baptist History.* Nashville, Tenn.: Historical Commission of the Southern Baptist Convention; Southern Baptist Historical Society, 1984. 64 p.

Seven essays on the history of church music in the Baptist Church. Published as a separately titled issue of the journal *Baptist History and Heritage*, 19/1

(Jan. 1984). Essays: "The Musical Heritage of Baptists" (J. P. Newport); "Turning Points in the Story of Baptist Church Music" (H. T. McElrath); "The Sunday School Board and Baptist Church Music" (W. L. Forbis); "Southern Baptist Contributions to Hymnody" (H. L. Eskew); "Music in Southern Baptist Evangelism" (D. W. Music); "Instrumental Music in Southern Baptist Life" (A. J. King); and "The Graded Choir Movement among Southern Baptists" (W. J. Reynolds). Documented with endnotes.

250. Eskew, Harry, David W. Music, and Paul A. Richardson. *Singing Baptists: Studies in Baptist Hymnody in America.* Nashville, Tenn.: Church Street Press, 1994. xii, 224 p. ISBN 0–8054–9824–9 ML3160.E85 S5 1994

A collection of articles about music in U.S. Baptist churches organized under four headings: (1) "Early New England Developments," (2) "Pastor-Hymnists of the 19th-Century South," (3) "Singing-School Tunebooks of the 19th-Century South," and (4) "Southern Baptist Hymnody." Articles include: "William Walker and his *Southern Harmony,*" "William Walker's *Southern Harmony:* Its Basic Editions," "*Christian Harmony* Singing in Alabama: Its Adaptation and Survival," "Use and Influence of Hymnals in Southern Baptist Churches up to 1915," and "Southern Baptist Contributions to Hymnody" (Eskew); "Oliver Holden (1765–1844): An Early Baptist Composer in America," "The First American Baptist Tunebook," "Music in the First Baptist Church of Boston, Massachusetts, 1665–1820," "Starke Dupuy: Early Baptist Hymnal Compiler," "The Hymns of Richard Furman," and "J. R. Graves' *The Little Seraph* (1874): A Memphis Tunebook" (D. W. Music); "Andrew Broaddus and Hymnody," "Eleazar Clay's *Hymns and Spiritual Songs* (1793)," "Basil Manly, Jr.: Southern Baptist Pioneer in Hymnody," and "Eli Ball, of Virginia" (Richardson). Facsimiles, illustrations, and tables; annotated bibliography of more than eighty writings on Baptist hymnody; index.

251. Gregory, David Louis. "Southern Baptist Hymnals (1956, 1975, 1991) as Sourcebooks for Worship in Southern Baptist Churches." D.M.A. dissertation. Louisville, Ky.: Southern Baptist Theological Seminary, 1994. viii, 201 p.

Study of the correlation between the history of the Southern Baptist Convention and publication of *Baptist Hymnal* (1956), *Baptist Hymnal* (1975), and *The Baptist Hymnal* (1991). Bibliography of approximately 180 writings, eleven hymnals and tunebooks, and ten interviews, conversations, and questionnaires; index.

252. Hitt, Gwen Keys. *We Shall Come Rejoicing: A History of Baptist Church Music in Mississippi.* Preface by Dan C. Hall. Starkville: Mississippi State University Press, 1984. 122 p. ML3160 .H57

Not intended to be a complete music history of Baptist church music in Mississippi. Covers significant events from 1780 to 1984. Photos and facsimiles; chronology.

253. McCommon, Paul. "Our Music Tradition" in *Music in the Worship Experience.* Compiled by William M. Anderson, Jr. Nashville, Tenn.: Convention Press, 1984. pp. 26–32.

Music in the Worship Experience is designed as a text for the Church Study Course of the Baptist Sunday School Board. The text consists of six chapters, each by a different author; McCommon's contribution is Chapter 3 which briefly recounts the history of Baptist church musical practices in England during the seventeenth and eighteenth centuries and in colonial America to present times. Illustrations and photographs.

254. Measels, Donald Clark. "A Catalog of Source Readings in Southern Baptist Church Music: 1828–1890." D.M.A. dissertation. Louisville, Ky.: Southern Baptist Theological Seminary, 1986. 2 vols.: vi, 146 p.; ix, 415 p.

"Writings on or about use of music by Baptists in the Southeast as described in selected newspapers beginning with origin of the papers (1828) to 1890." Bibliography in three sections: (1) "Pre-Convention (1828–1845)," (2) "Formation of Convention-Civil War (1846–1865)," and (3) "Post Civil War (1866–1890)." Volume 2 presents reprints of selected writings. Bibliography of thirty-nine writings in addition to those chosen for the catalog.

255. Murrell, Irvin Henry. "An Examination of Southern Ante-Bellum Baptist Hymnals and Tunebooks as Indicators of the Congregational Hymn and Tune Repertories of the Period with an Analysis of Representative Tunes." D.M.A. dissertation. La.: New Orleans Baptist Theological Seminary, 1984. ix, 131 p.

Overview of pre–Civil War Southern Baptist hymnody. Identifies common hymns and tunes that appear in at least two-thirds of Southern antebellum Baptist hymnals and tunebooks. Examines common characteristics of the common tune repertory. Musical examples; several useful appendixes, including brief biographical information about fifty-four compilers of hymnals and tunebooks; bibliography of approximately forty-one writings.

256. Reynolds, William Jensen. *Baptist Hymnody in America: From Roger Williams to Samuel Francis Smith.* Introduction by Gordon L. Borror. Portland, Ore.: Western Conservative Baptist Seminary, 1981. 73, iv p.

Lecture presented at Western Conservative Baptist Seminary, May, 1981. Despite title, actually covers Baptist music practices from the founding of a Baptist church in Providence, Rhode Island, organized by Roger Williams in 1639 to Samuel Francis Smith's and Baron Stow's *The Psalmist* published in 1843 to hymnody in the twentieth century. Discusses music of Rhode Island Baptist churches in early seventeenth century, music of Welsh Baptists in New York in early eighteenth century, various significant hymnals and tunebooks, spiritual songs, camp meeting songs, the influences of Isaac Watts, John Rippon, and John Leland, biblical imagery, and contemporary trends in music practices. Documented with endnotes.

257. Sellers, Mary Josephine. "The Role of the Fine Arts in the Culture of Southern Baptist Churches." Ph.D. dissertation. New York: Syracuse University, 1968. xiii, 336 p.

Examines four categories of the fine arts within Southern Baptist churches: church architecture, hymnody, pictorial art, and fictional literature and drama. Discussion on hymnody covers congregational singing, role of choirs and musical instruments, the "Great Awakening" movement, and significant hymnals and music collections. Includes a bibliography of fifty-one favorite songs of Southern Baptists. Facsimiles; general bibliography of approximately three hundred writings.

258. Titon, Jeff Todd. *Powerhouse for God: Speech, Chant, and Song in an Appalachian Baptist Church.* Austin: University of Texas Press, 1988. xviii, 523 p. ISBN 0–292–76485–5 BX6480.S8434 T57 1988

Powerhouse for God: Sacred Speech, Chant, and Song in an Appalachian Baptist Church. Recorded by Jeff Todd Titon and Kenneth M. George. Chapel Hill: University of North Carolina Press, 1982. 2 sound discs (33 1/3 rpm, 12 in.) or 2 audio cassettes. ISBN 0–8078–4084-X

Explains religious practices of the Fellowship Independent Baptist Church near Stanley, Virginia. One chapter, "Singing," addresses the social organization of the music, repertory, and ideas about music. Musical examples, illustrations, maps, photographs, and tables; bibliography of about 250 interviews and writings; expansive index. Accompanied by a recording of representative music and spoken dialogue.

See also: Burrage, Henry S. *Baptist Hymn Writers and Their Hymns* (item 35); Duncan, Curtis Daniel. "A Historical Survey of the Development of the Black Baptist Church in the United States and a Study of Performance Practices Associated with Dr. Watts Hymn Singing: A Source Book for Teachers" (item 228)

BRETHREN

259. Church of the Brethren. Parish Ministries Commission. *We Gather Together: Worship Resources for the Church of the Brethren.* Elgin, Ill.: Brethren Press, 1979. viii, 236 p.

Book of worship resources for the Church of the Brethren. Sections 4 through 8 deal with congregational singing, choral music, instrumental music, children's choirs, and the use of sound recordings and media. Musical examples, photographs, and tables; several classified bibliographies of music and readings throughout; glossary of terms.

260. Hinks, Donald R. *Brethren Hymn Books and Hymnals, 1720–1884.* Gettysburg, Pa.: Brethren Heritage Press, 1986. 205 p. ML3161 .H56 1986

A history of Brethren hymn books and hymnals in Europe and the United States from the early eighteenth century to the late nineteenth century. Includes an annotated bibliography of approximately 180 major Brethren hymn books and hymnals and 9 Brethren-related hymn books published between 1720 and 1884. Facsimiles.

261. Martin, Betty Jean. "The Ephrata Cloister and its Music, 1732–1785: The Cultural, Religious, and Bibliographical Background." Ph.D. dissertation. College Park: University of Maryland, 1974. vii, 385 p.

In three parts: (1) historical and religious survey of the Ephrata Cloister, an eighteenth-century monastic settlement in Lancaster County, Pennsylvania; (2) study of the cloister's hymnals and manuscripts located in the Library of Congress, the Historical Society of Pennsylvania, the Library Company of Philadelphia, the New York Public Library, and the Seventh Day Baptist Historical Society at Plainfield, New Jersey; and (3) analysis of the hymn texts and original music. Musical examples; bibliography of approximately 160 writings.

262. Sachse, Julius Friedrich. *The Music of the Ephrata Cloister; Also, Conrad Beissel's Treatise on Music as Set Forth in a Preface to the "Turtel Taube" of 1747, Amplified with Fac-Simile Reproductions of Parts of the Text and Some Original Ephrata Music of the Weyrauchs Hügel, 1739; Rosen und Lilien, 1745; Turtel Taube, 1747; Choral Buch, 1754, etc.* Lancaster, Pa.: Julius Friedrich Sachse, 1903. 108 p. ML200.8.E64 C64 1903; reprint, New York: AMS Press, 1971. ISBN 0–404–05500–1 ML200.8.E64 C64 1971

Pennsylvania-German music of the Ephrata Cloister (Church of the Brethren). Originally published in 1903 as a reprint from volume XII of the "Proceedings of the Pennsylvania-German Society." Includes Conrad Beissel's treatise on music. Photographs and numerous facsimiles; expansive index.

See also: Faus, Nancy Rosenberger, ed. *The Importance of Music in Worship* (item 333)

CATHOLIC

263. Boccardi, Donald. *The History of American Catholic Hymnals: Since Vatican II.* Chicago.: G. I. A. Publications, 2001. xv, 169 p. ISBN 1–57999–121–1 BV360.A1 B63 2001

Historical survey of the development of Catholic hymnals in the United States during the latter half of the twentieth century. Bibliography of more than two hundred writings, a few audio/visual resources, and fifty Catholic and ten Protestant hymnals; expansive general index.

264. Catholic Church. National Conference of Catholic Bishops. Bishops' Committee on the Liturgy. *Liturgical Music Today: A Statement of the Bishops' Committee on the Liturgy on the Occasion of the Tenth Anniversary of Music in Catholic Worship.* Washington, D.C.: Bishops' Committee on the Liturgy, National Conference of Catholic Bishops, 1982. 28 p. ML3007 .C45 1982

Published ten years after *Music in Catholic Worship* (1972), a statement of the Bishops' Committee on the Liturgy. Presents the Bishops' views on musical topics, including: the place, function, and form of song; language and musical idioms; music in the Eucharist, Liturgy of the Hours, and celebration of various sacraments and rites, namely Christian initiation, reconciliation, marriage, and burial; instrumental music; and other music topics. Documented with endnotes.

265. Catholic Church. National Conference of Catholic Bishops. Bishops' Committee on the Liturgy. *Music in Catholic Worship.* Rev. ed. Washington, D.C.: Bishops' Committee on the Liturgy, National Conference of Catholic Bishops, 1983. 31 p. ML3011 .C35 1983

Revision of the 1972 edition. Statement of the Bishops' Committee on the Liturgy. Documented with endnotes. Also published under the title *The Music Documents: Music in Catholic Worship & Liturgical Music Today* (Oregon Catholic Press, 1995).

266. Damian, Ronald. "A Historical Study of the Caecilian Movement in the United States." D.M.A. dissertation. Washington, D.C.: Catholic University of America, 1984. vi, 300 p.

Traces the Caecilian movement in the United States, 1874 to 1965. Music of the Caecilian movement is characterized by a return to the a cappella style of the sixteenth century and to the use of Gregorian chant. Musical examples, photocopied photographs, and facsimiles of *Caecilia* magazine covers. Bibliography of more than two hundred writings.

267. Day, Thomas. *Where Have You Gone, Michelangelo?: The Loss of Soul in Catholic Culture.* New York: Crossroad, 1993. xii, 240 p. ISBN 0–8245–1396–7 BX1406.2 .D38 1993

     Very personal view of Catholic worship practices in the United States. Chapter 4, "Depressing Music: It Doesn't Belong," expresses the author's concerns regarding current Catholic music. Documented with endnotes; expansive index.

268. Day, Thomas. *Why Catholics Can't Sing: The Culture of Catholicism and the Triumph of Bad Taste.* New York: Crossroad, 1990. viii, 183 p. ISBN 0–8245–1035–6 BX1406.2 .D39 1990

     Addresses the effect of pop culture on Catholic church music. Based on personal experiences. Documented with endnotes; expansive index.

269. Deiss, Lucien. *Visions of Liturgy and Music for a New Century.* Translated by Jane M.-A. Burton. English text edited by Donald Molloy. Collegeville, Minn.: Liturgical Press, 1996. xi, 242 p. ISBN 0–8146–2298–4 ML3080 .D45 1996

     English translation of *Vision de la Musique et de la Liturgique.* Considers sacred music in the Catholic liturgy, especially in the Eucharistic celebration. In three parts: (1) the ministerial function of music and song; (2) participants involved in the music of the liturgy; and (3) examines different chants of the Eucharistic celebrations and classifies each according to its genre. Musical examples; documented with footnotes; expansive index.

270. DeSanctis, Michael E. "Some Artistic Aspects of Catholic Liturgical Reform: A Comparative Study of the Influence of Vatican Council II on Music and Architecture for the Liturgy." Ph.D. dissertation. Athens: Ohio University, 1985. xxiv, 263 p.

     Historical survey of the Liturgical Movement and its influence on Catholic sacred music in the United States prior to and following Vatican Council II (1962–1965). Analyzes selected examples of musical settings of the Ordinary of the Mass. Musical examples, illustrations, and photographs; classified bibliography (Vatican documents; American Episcopal documents; general; Vatican Council II; liturgy; music; architecture) of approximately 270 writings.

271. Dunne, John S. *The Music of Time: Words and Music and Spiritual Friendship.* Indiana: University of Notre Dame Press, 1996. xii, 226 p. ISBN 0–268–01423-X BX2350.2 .D87 1996

     Examines religious aspects of Catholic church music. Documented with endnotes; expansive index.

272. Fellerer, Karl Gustav. *The History of Catholic Church Music.* Translated by Francis A. Brunner. 2nd ed. Baltimore: Helicon Press, 1961. 235 p. ML3002 .F32 1961; reprint, Westport, Conn.: Greenwood Press, 1979. ISBN 0–313–21147–7 ML3002 .F32 1979

English translation of the second edition of Fellerer's *Geschichte der katholischen Kirchenmusik* (1949) with additional notes and corrections supplied by Fellerer and new information on church music in America supplied by Joan Boucher. Boucher's contribution is small, but valuable. The general index is helpful in identifying subjects related to church music in America. Classified bibliography of forty items; composer and general indexes.

273. Foley, Edward, and Mary E. McGann. *Music and the Eucharistic Prayer.* Washington, D.C.: Pastoral Press, 1988. 52 p. ISBN 0–912405–50–3 BX2015.6 .F644 1988; reprint, Collegeville, Minn.: Liturgical Press, 1988. ISBN 0–8146–1934–7 BX2015.6 .F644 1988b

Examines major developments in American liturgical music within the Catholic Church since publication of *The Constitution on the Sacred Liturgy* (1963) by the Vatican Council II (1962–1965). Focuses on the relationship between music and the Eucharistic prayer. Musical examples; documented with endnotes.

274. Foley, Edward. *Ritual Music: Studies in Liturgical Musicology.* Beltsville, Md.: Pastoral Press, 1995. vi, 201 p. ISBN 1–56929–057–1 ML3000 .F65 1995

Collection of eight essays on church music in the Catholic Church. Essays: "Judaeo-Christian Ritual Music: A Bibliographic Introduction to the Field;" "The Auditory Environment of Emerging Christian Worship;" "The Cantor in Historical Perspective;" "Martin Luther: A Model Pastoral Musician;" "Towards a Sound Theology;" "From *Music in Catholic Worship* to the 'Milwaukee Document';" "Musical Forms, Referential Meaning, and Belief;" and "The Evaluation of Roman Catholic Ritual Music: From Displacement to Convergence." Musical examples; documented with endnotes; general, expansive index.

275. Funk, Virgil C., ed. *Initiation and Its Seasons.* Washington, D.C.: Pastoral Press, 1990. vi, 143 p. ISBN 0–912405–74–0 ML3002 .P25 1981 v.3

Collection of fourteen essays by various authors. Essays relating to music: "Choosing Music for the Rites" (D. Hollier); "Liturgical Glue" (D. Haas); "Music for the Three Days" (R. Strusinski); and "Initiation at Easter: Challenging the Musician" (R. N. Fragomeni).

276. Funk, Virgil C., ed. *The Pastoral Musician.* Washington, D.C.: Pastoral Press, 1990. vi, 120 p. ISBN 0–912405–76–7 ML3002 .P25 1981 v.5

Collection of ten essays by various authors: "The Musician as Minister" (N. Mitchell); "Lay Ministry and Lay Musicians" (V. S. Finn); "Collaborative Ministry: Do We Want It?" (B. Fleo); "The Musician: Transformed through Excellence" (C. Serjak); "The Mind-Set of a Musical Minister" (D. E. Pilarczyk); "The Importance of Prayer for the Musician" (J. Gelineau); "Describing the Pastoral Musician's Role" (C. Conley); "Directives to a Pastoral Musician" (M. T. Kalb); "Seven Rules for Making It Work" (R. Strusinski); and "The Care and Feeding of Pastoral Musicians" (J. Ferguson).

277. Funk, Virgil C., ed. *The Singing Assembly.* Washington, D.C.: Pastoral Press, 1991. vi, 115 p. ISBN 0–912405–80–5 ML3002 .P25 1981 v.6

Collection of fifteen essays by various authors: "Assembly: Remembering into the Future" (M. Searle); "Full, Conscious, and Active Participation Is for All the People" (J. Gurrieri); "The Art of Assemblying" (T. Guzie); "Do It with Style" (Funk); "Participation: Is It Worth the Effort?" (L. Deiss); "The Congregation's Active Participation Is Performance" (L. Madden); "Two Become One: Performance and Participation" (P. Philibert); "Balancing Performance and Participation" (J. Gelineau); "For Congregational Song, Prayer Is First" (M. A. Piil); "Helping Your Congregation to Participate" (C. J. McNaspy); "Congregational Singing: What? How? Why? . . . But!" (F. Moleck); "On Your Mark! Get Set! Ten Steps" (F. Brownstead); "How Can We Keep Them Singing?" (G. Truitt); "Ten Commandments for Those Who Love the Sound of a Singing Congregation?" (C. R. Gardner); and "Claim Your Art" (R. Weakland).

278. Funk, Virgil C., ed. *Sung Liturgy: Toward 2000 A.D.* Washington, D.C.: Pastoral Press, 1991. vi, 111 p. ISBN 0–912405–80–5 ML3007 .S86 1991

Six essays on liturgy and music in today's Catholic Church. Essays dealing with music are: "Music at the Crossroads: Liturgy and Culture" (E. Costa); "Theater, Concert, or Liturgy: What Difference Does It Make?" (E. Foley); and "The Future of Church Music" (Funk). Documented with endnotes.

279. Funk, Virgil C., and Gabe Huck, eds. *Pastoral Music in Practice.* Washington, D.C.: National Association of Pastoral Musicians; Chicago: Liturgy Training Publications, 1981. vi, 152 p. ISBN 0–9602378–3–6 ML3002 .P25 1981 v.1

Collection of nineteen articles previously published in *Pastoral Music* between 1976 and 1979. Articles include: "Music Ministry, Today and Tomorrow" (R. Weakland); "The Dilemma of Pastoral Music" (J. Melloh);

"The Musician as Minister" (N. Mitchell); "The Animator" (J. Gelineau); "The Church as a Community of Prayer" (G. Diekmann); "A God Who Hears" and "Six Minor Heresies in Today's Music" (Mitchell); "Two Become One: Performance and Participation" (P. Philibert); "Choosing Music: No Small Task" (T. Conry); "Setting the Tone" (R. Dufford); "Beyond the Spectator Sacraments" (K. Meltz); "Pastoral Liturgy is NOT in the Book" (R. Keifer); "Musical, Liturgical, Pastoral Judgments: New Songs, New Judgments?" (W. Bauman); "Our People Just Don't Like to Sing," "Good Environment Helps," "New Music: Step by Step," and "How the Organist Can Lead the Congregation" (R. J. Batastini); "Prayer and Music: Singing the Meaning of Words" (J. M. Joncas); and "Hymns in History" (A. Parker).

280. Grimes, Robert R. *How Shall We Sing in a Foreign Land?: Music of Irish-Catholic Immigrants in the Antebellum United States.* Notre Dame, Ind.: University of Notre Dame Press, 1996 (rep. 1998). xi, 237 p. ISBN 0–268–01110–9 ML3554 .G75 1996

Originated from author's doctoral dissertation (University of Pittsburgh, 1992). Study of Irish Catholic musical life in the United States between 1830 and 1860. Musical examples, tables, and an illustration; bibliography of six manuscript sources, twenty-six nineteenth-century newspapers and periodicals, and approximately 180 writings; expansive index.

281. Hayburn, Robert F. *Papal Legislation on Sacred Music, 95 A.D. to 1977 A.D.* Collegeville, Minn.: Liturgical Press, 1979. xiv, 619 p. ISBN 0–8146–1012–9 ML3002 .H43

Compilation of writings on sacred music by various popes, translated into English where needed. Photographs and facsimiles; various appendixes; bibliography of about 550 writings; expansive index.

282. Higginson, J. Vincent. *History of American Catholic Hymnals: Survey and Background.* Springfield, Ohio: Hymn Society of America, 1982. xi, 286 p. BV360.A1 H53 1982

Survey of more than one hundred Catholic hymnals in the vernacular in the United States from 1787 to 1975. Documented with endnotes; expansive index.

283. Joncas, Jan Michael. *From Sacred Song to Ritual Music: Twentieth-Century Understandings of Roman Catholic Worship Music.* Collegeville, Minn.: Liturgical Press, 1997. viii, 115 p. ISBN 0–8146–2352–2 ML3007 .J66 1997

Nine papal, conciliar, curial, bishop's conference, and scholar's documents examined in response to five questions: (1) What is Roman Catholic worship music?; (2) What is the purpose of Roman Catholic worship music?; (3) What qualities should Roman Catholic worship music exhibit?; (4) What people are to make Roman Catholic worship music?; and (5) What instruments are to make Roman Catholic worship music? Documented with endnotes; no index.

284. Licon, Peggy Ann. "Twentieth-Century Liturgical Reform in the Catholic Church and a Sample of Current Choral Literature (1989)." D.M.A. dissertation. Tempe: Arizona State University, 1989. 2 vols.: xi, 535 p.

The effect of Vatican Council II (1962–1965) upon musical development within the Catholic Church in the United States. Includes an annotated bibliography of nearly ninety representative, sacred liturgical choral works for mixed, treble, men's, and double chorus currently used in U.S. cathedrals. Provides composer/arranger, title, publishing information, voicing, accompaniment, text, vocal ranges, liturgical season, music period, composer's dates, and remarks about harmony, tempo, meter, rhythm, melody, texture, articulation, form, dynamics, and timbre. Tables; bibliography of approximately two hundred writings.

285. McGann, Mary E. *Exploring Music as Worship and Theology: Research in Liturgical Practice.* Collegeville, Minn.: Liturgical Press, 2002. 81 p. ISBN 0–8146–2824–9 ML3080 .M34 2002

Study of music as liturgical practice within the Catholic Church.

286. McGann, Mary E. "Interpreting the Ritual Role of Music in Christian Liturgical Practice." Ph.D. dissertation. Berkeley, Calif.: Graduate Theological Union, 1996. xi, 282 p.

The role of music in a Catholic parish (Our Lady of Lourdes in San Francisco, California) which embraces African-American music traditions. Explores style and patterns of worship, musical leadership, and liturgical repertoire, especially gospel music. Musical examples, illustrations, and tables; bibliography of approximately 230 writings and a few hymnals.

287. Milwaukee Symposia for Church Composers. *The Milwaukee Symposia for Church Composers: A Ten-Year Report, July 9, 1992.* Washington, D.C.: Pastoral Press; Chicago: Liturgy Training Publications, 1992. 32 p. ISBN 0–912405–43–0 (PP), ISBN 0–929650–91–3 (LTP) ML3011 .M55 1992

"A report on ten years of observation, study, reflection and dialogue concerning the nature and quality of liturgical music in the United States, especially within the Roman Catholic tradition." Addresses music as a language of faith, liturgical formation, liturgical preparation, liturgical and musical structures, textural considerations, cross-cultural music making, models of musical leadership, technology and worship, and the musical-liturgical-pastoral judgment. Documented with endnotes.

288. Nemmers, Erwin Esser. *Twenty Centuries of Catholic Church Music.* Milwaukee, Wis.: Bruce Publishing, 1949. xvii, 213 p. ML3002 .N35; reprint, Westport, Conn.: Greenwood Press, 1978. ISBN 0–313–20542–6 ML3002 .N35 1978

    Chapter 6, "History of American Catholic Church Music," briefly surveys Catholic church music in North America, including Mexico, Puerto Rico, the United States, and Canada from the sixteenth century to the mid-twentieth century. For Chapter 6, a bibliography of about forty writings and a bibliography of twenty-three collections of music published in the United States between 1787 and 1860. Musical examples, illustrations, and photos; expansive index.

289. Plank, Steven Eric. *"The Way to Heavens Doore": An Introduction to Liturgical Process and Musical Style.* Metuchen, N.J.: Scarecrow Press, 1994. xiv, 183 p. ISBN 0–8108–2953–3 ML3000 .P53 1994

    Explores the relationship between musical style and liturgical processes. Primary focus on the Catholic Mass and the daily Office. Musical examples; bibliography of more than 120 writings; general index.

290. Price, Milburn. "The Impact of Popular Culture on Congregational Song." *The Hymn: A Journal of Congregational Song,* 44/1 (Jan. 1993): 11–19

    Reviews the influence of popular culture on music within the Catholic Church since Vatican Council II (1962–1965). Illustrations; documented with endnotes; bibliography of thirty-three illustrative hymns and choruses.

291. Ray, Mary Dominic, and Joseph H. Engbeck. *Gloria Dei: The Story of California Mission Music.* Sacramento: State of California, Department of Parks and Recreation, 1974; reprint, Berkeley: University of California, 1975. 24 p. ML3011.R3 G5x 1975

    Attractive publication with numerous illustrations, photos, and facsimiles. California mission music of the eighteenth and nineteenth centuries, specifically mission bells, padre musicians, mission choirs, music and instruments of the California Native Americans, sacred and secular musical repertoire, and the role of barrel organs in mission music. Bibliography of eight writings.

292. Savaglio, Paula Clare. "Polish-American Music in Detroit: Negotiating Ethnic Boundaries." Ph.D. dissertation. University of Illinois at Urbana-Champaign, 1992. vii, 295 p.

    Covers popular music as well as church music. Separate chapter on church music titled "Music in Polish Churches in the Roman Catholic Archdiocese of Detroit." Discusses music of Polish-language and English-language

Masses, including characteristics of hymn repertoire, role of the choir, and the polka Mass. Musical examples and illustrations; bibliography of twenty-one interviews, nearly two hundred writings, and a discography of approximately forty-five recordings (mostly popular music).

293. Silverberg, Ann Louise. "Cecilian Reform in Baltimore, 1868–1903." Ph.D. dissertation. University of Illinois at Urbana-Champaign, 1992. xvii, 565 p.

Catholic liturgical music reform in late-nineteenth- through early-twentieth-century Baltimore, Maryland. Musical examples and tables; bibliography of approximately 141 writings.

294. Winter, Miriam Therese. *Why Sing?: Toward a Theology of Catholic Church Music.* Washington, D.C.: Pastoral Press, 1984. 346 p. ISBN 0–912405–07–4 ML3010 .W55 1984

Based on the author's doctoral dissertation, "Vatican II in the Development of Criteria for the Use of Music in the Liturgy of the Roman Catholic Church in the United States and Their Theological Bases" (Princeton Theological Seminary, 1983). Purpose: "to establish a theological context for contemporary discussions on Catholic Church music." Bibliography of more than twenty-one hundred writings.

See also: Funk, Virgil C., ed. *Children, Liturgy, and Music* (item 580)

CHRISTIAN (DISCIPLES OF CHRIST)

295. Helseth, David C. "The Changing Paradigm of Congregational Music: A Disciples of Christ Response." D.Min. dissertation. Calif.: Claremont School of Theology, 1997. v, 210 p.

Examines the increasing use of contemporary hymns within services of the Christian Church (Disciples of Christ). Aims "to show that the many contemporary hymns being written and sung today can help revitalize the worship of mainline congregations without [discarding] the traditional order of service." Musical examples and tables; bibliography of 123 writings and twenty-four musical scores.

CHRISTIAN REFORMED

296. Polman, Bertus Frederick. "Church Music & Liturgy in the Christian Reformed Church of North America." Ph.D. dissertation. Minneapolis: University of Minnesota, 1980. iii, 260 p.

Surveys sixteenth-century Dutch psalmody, liturgy influences, and later developments in North America from the mid-1800s through the 1970s. Musical examples and tables; classified bibliography of more than 180 writings and six psalters and hymnals.

CHURCH OF CHRIST

297. Colley, Gary, ed. *Music in New Testament Worship: The Fourteenth Annual Southwest Lectures, April 9–12, 1995.* Austin, Tex.: Southwest Publications, 1995. 370 p.

Collection of twenty-four lectures delivered at the 1995 Southwest Lectures, conducted by the Southwest Church of Christ, Austin, Texas. Lectures: "Establishing Bible Authority" (P. B. Cotham); "Proper Attitude toward Worship" (Colley); "God's Design in Singing" (J. Meador); "The Divine Command of the New Testament to Sing" (G. Hitchcock); "Implications of the Divine Command to Sing in the New Testament" (Al Macias); "An In-Depth Study of Colossians 3:16–17" (J. Gilmore); "An In-Depth Study of Ephesians 5:19" (J. Moore); "Instrumental Music in Christian Worship" (A. O. Colley); "Instrumental Music in the Old Testament" (Gilmore); "The Absence of Instrumental Music in Temple Worship (Implications)" (G. Elkins); "The Kind of Music in the Early Church" (Hitchcock); "The Restoration Movement—Valid Today?" (R. R. Taylor, Jr.); "Review of the Instrumental Music Debates of Foy E. Wallace, Jr." (N. Patterson); "Review of Hardeman-Boswell Debate on Music" (B. Lockwood); "Review of M. C. Kurfee's Book on Instrumental Music" (B. Dobbs); "Are Solos, Quartets, and Choirs Scriptural?" (Dobbs); "What about Harps in Heaven?" (Elkins); "What about Vocal Bands?" (L. Mitchell); "Instrumental Music in the Home—Why Not the Church?" (R. Williamson); "Is Instrumental Music an Aid?" (D. McClish); "Review of Old Testament Passages on Music in Worship" (P. Sain); "Review of New Testament Passages on Music in Worship" (B. Lockwood); "The Influence of One" (I. Taylor); and "The Christian's Garden of Contentment" (Taylor).

298. Ferguson, Everett. *A Cappella Music in the Public Worship of the Church.* 3rd ed. Fort Worth, Tex.: Star Bible Publications, 1999. 93 p. ISBN 1–56794–217–2 ML3001 .F36 1999

The author, from a Church of Christ liturgical background, advocates a cappella music in public worship based on: "(1) an analysis of the New Testament evidence, (2) a testing of one's interpretation of the New Testament by the testimony of church history, and (3) a consideration whether there are doctrinal or theological reasons which explain or give meaning to the Biblical and historical evidence." Glossary of terms; bibliography of nearly fifty writings by ancient authors and a bibliography of ten writings by contemporary authors.

299. Jackson, James L. "Music Practices among Churches of Christ in the United States, 1970." D.Mus.Ed. dissertation. Norman: University of Oklahoma, 1970. viii, 121 p.

Music practices of Churches of Christ in the United States in the year 1970. Stated purposes: "(1) to determine what hymnals are being used, what hymns are used most frequently, and the qualifications of the song leaders; (2) to compare practices among congregations in small communities with those in larger cities; (3) to compare practices among different regions of the nation; (4) to evaluate the music practices of the churches; and (5) to make recommendations for future improvements and developments." Musical examples and tables; bibliography of 19 writings and 12 hymnals.

CHURCH OF CHRIST, SCIENTIST

300. Robertson, Patricia Lou. "The Role of Singing in the Christian Science Church: The Forming of a Tradition." Ph.D. dissertation. New York University, 1996. viii, 353 p.

Past and present musical practices in the Church of Christ, Scientist (founded 1879), also known as the Christian Science Church. Surveys development of the church and discusses the views of its founder Mary Baker Eddy (1821–1910). The music used in the late nineteenth century had roots in the Protestant musical heritage, using choirs and soloists. Use of choirs was abolished in 1898. Today, music in the church includes hymns and vocal solos, borrowing from Protestant musical traditions. Musical examples, tables, and photocopied facsimiles; bibliography of approximately 175 writings; several appendixes.

CHURCH OF GOD

See: Buehler, Kathleen D. *Heavenly Song: Stories of Church of God Song Writers and Their Songs* (item 34)

CHURCH OF GOD IN CHRIST

301. Neely, Thomasina. "Belief, Ritual, and Performance in a Black Pentecostal Church: The Musical Heritage of the Church of God in Christ." Ph.D. dissertation. Bloomington: Indiana University, 1993. xxiii, 553 p.

Analysis of "traditional and modern forms of black sacred music" found in the Church of God in Christ. Study "focuses on the ways in which Pentecostal belief is represented and organized through music and ritual behaviors." Musical examples; bibliography of more than 170 writings and twenty-five interviews; discography of thirty-five recordings.

EPISCOPAL/ANGLICAN

302. Clark, Linda J. *Music in Churches: Nourishing Your Congregation's Musical Life.* Bethesda, Md.: Alban Institute, 1994 (rep. 1995). xv, 121 p. ISBN 1–56699–134-X ML3001 .C58 1994

Examines musical practices in Episcopal and United Methodist churches in Massachusetts, Rhode Island, and Connecticut. Tables; bibliography of approximately sixty writings.

303. Criswell, Paul Douglas. "The Episcopal Choir School and Choir of Men and Boys in the United States: Its Anglican Tradition, Its American Past and Present." Ph.D. dissertation. College Park: University of Maryland, 1987. iv, 161 p.

Traces the development of the Episcopal Choir School in the United States, a descendent of the Anglican Choir School associated with the Church of England. Bibliography of nearly sixty writings.

304. Doran, Carol, and William H. Petersen. *A History of Music in the Episcopal Church.* 2nd ed. S.1.: Association of Anglican Musicians Continuing Education Project for Clergy and Musicians, 1991. 166 p. ML3166 .D673 1991

Originally published in 1984. Study organized into six units. Most units pertain to the history of the church in England. The latter three units—"English Christianity Comes to the New World," "The Nineteenth Century (1789–1914)," and "The Present Age," respectively—include discussion of Episcopal church music in the United States. Music examples and facsimiles; bibliographies of approximately fifteen to thirty writings at the end of each unit; no index.

305. Glover, Raymond F., ed. *The Hymnal 1982 Companion.* New York: Church Hymnal Corporation, 1990. 4 vols.: xvi, 764 p.; xvi, 749 p.; xii, 718 p.; 719–1392 p. ISBN 0–89869–143–5 ML3166 .H88 1990

Excellent resource. Four volumes with a total of nearly 3,000 pages: Vol. 1: essays on church music; Vol. 2: service music and biographies; Vol. 3A: hymns 1–384; and Vol. 3B: hymns 385–720. A companion to *The Hymnal 1982* of the Episcopal Church, though the scope of work projects beyond its initial purpose, serving as valuable source about composers and providing analytical and historical information about hymns used by Episcopal churches and other denominations.

Vol. 1: Selected essays on church music in the United States: "What is Congregational Song?" (Glover); "The Spirituality of Anglican Hymnody: A Twentieth-Century American Perspective" (C. P. Daw, Jr.); "Popular Religious Song" (C. Doran); "Cultural Diversity" (H. C. Boyer) is primarily concerned with white and African-American religious cultures; "Accompaniment and Leadership of Congregational Song" (Doran); "The Publication of the Hymnal of the Episcopal Church" (L. L. Ellinwood, C. G. Manns); "The Creation of *The Hymnal 1982*" (Glover); "Two Hundred Years of Service Music in Episcopal Hymnals" (J. H. Litton); "A Survey of Service Music in *The Hymnal 1982*" (E. Morris Downie); "Charles Winfred Douglas and Adaptation of Plainsong to English Words in the United States" (B. E. Ford); "Harmonized Chant" (R. M. Wilson); "Contemporary Use of Chant" (R. Proulx, Glover); "English Metrical Psalmody" (R. A. Leaver) discusses metrical psalmody in England and the United States and Puritan psalters in New England; "Psalmody in America to the Civil War" (Manns); "Hymnody in the United States from the Civil War to World War I (1860–1916)" (P. Westermeyer); "Protestant Hymn-Singing in the United States, 1916–1943: Affirming an Ecumenical Heritage" (D. Farr); "Hymnody in the United States since 1950" (R. Schulz-Widmar); and "A Checklist of Music Publications for the Protestant Episcopal Church in the United States of America to 1940" (Manns). Glossary of terms; general bibliography of more than three hundred writings with approximately sixty on American hymnody; expansive general index.

Vol. 2: Discussion of text, music, and historical background for each of the 449 works intended for service music included in *The Hymnal 1982.* Also, nearly eight hundred biographies of composers whose works are included in the hymnal. Glossary of terms; bibliography of approximately seventy liturgical commentaries, handbooks, dictionaries, and studies of particular services and approximately forty-five collections and editions of music; indexes to *The Hymnal 1982* for (1) composers, arrangers, and sources of service music, and (2) titles.

Vols. 3A–3B: Discussion of each hymn in *The Hymnal 1982,* covering music and text. Many musical examples; glossary of terms; separate indexes for: authors, translators, and sources; composers, arrangers, and sources for hymns; tune names; first lines; and metrical index with incipits.

306. Leist, Stephen G. *The History of Music at Christ Church Cathedral: 1796–1996.* Lexington, Ky.: Christ Church Episcopal Cathedral, 1998. v, 100 p. ML3166 .L45 1998

Specialized history of music in Christ Church Cathedral, Lexington, Kentucky, included here because of its two-hundred-year historical scope. Introduced by synopsis of Western church music from the medieval period to the present. Illustrations, photos; bibliography of more than thirty writings, interviews, and church records.

307. Riedel, Jane Rasmussen. *Musical Taste as a Religious Question in Nineteenth-Century America.* Lewiston, N.Y.: Mellen Press, 1986. xxvi, 603 p. ISBN 0–88946–664–5 ML3166 .R37 1986

Examines documents concerning music in the Episcopal Church in the United States, 1830s through 1860s. Selected reprints from the *Episcopal Magazine*, *Churchman*, *Gospel Messenger*, and *Church Journal*, among other writings, with ample discussion offered by Riedel. Writings primarily by clergy and laymen, rather than musicians.

308. Wilhite, Charles Stanford. "Eucharistic Music for the Anglican Church in England and the United States at Mid-Twentieth Century (1950–1965): A Stylistic Study with Historical Introduction." Ph.D dissertation. Iowa City: University of Iowa, 1968. x, 307 p.

Focused study on the Episcopal/Anglican Eucharist, variously referred to as the Lord's Supper, the Holy Communion, and the Mass. Discusses the nature and purpose of corporate worship and describes the development of the *Book of Common Prayer*; provides historical survey of Episcopal/Anglican church music in England and the United States; and presents historical, sociological, theological, and musical bases for critical evaluation of Eucharist music. Chapters 4 and 5: Episcopal/Anglican Eucharistic music in England and the United States, including a classified bibliography with analysis of about fifty-five works. Four appendixes: (1) classified bibliography of about forty additional musical works not discussed in detail in the text; (2) classified list of twenty-five canticles and eighty anthems; (3) annotated bibliography of nineteen historical writings and twenty-one historical collections and editions of music concerning Episcopal/Anglican church music; and (4) a discography of seven works discussed in Chapters 4 and 5. Musical examples; bibliography of about two hundred writings; index of works discussed in Chapters 4 and 5.

309. Wilson, Ruth Mack. *Anglican Chant and Chanting in England, Scotland, and America, 1660 to 1820.* Foreword by Nicholas Temperley. Oxford: Clarendon Press; New York: Oxford University Press, 1996. xix, 332 p. ISBN 0–19–816424–6 ML3166 .W55 1996

Originated in part as the author's doctoral dissertation, "Anglican Chant and Chanting in England and America, 1660–1811" (University of Illinois at Urbana-Champaign, 1988). Chapter 8, "Early Episcopal Music in America," covers Episcopal chant from the late eighteenth century to the early nineteenth century. Musical examples; general bibliography of approximately fifty music manuscripts and 270 writings relating to music in England, Scotland, and the United States; general expansive index.

See also: Alexander, J. Neil, ed. *With Ever Joyful Hearts: Essays on Liturgy and Music: Honoring Marion J. Hatchett* (item 424); Barr, Wayne Anthony. "The History of the Pipe Organ in Black Churches in the United States of America" (item 225); Campbell, Donald Perry. "Puritan Belief and Musical Practices in the Sixteenth, Seventeenth, and Eighteenth Centuries" (item 370); Jackson, Irene V., ed. *More Than Dancing: Essays on Afro-American Music and Musicians* (item 231)

EVANGELICAL LUTHERAN CHURCH IN AMERICA

310. Nordin, John P. "'They've Changed the Hymns!' or, An Investigation of Changes in Hymnody in the Hymnbooks of the Evangelical Lutheran Church in America." *The Hymn: A Journal of Congregational Song,* 47/1 (Jan. 1996): 25–33

Changes in Lutheran hymn singing in America observed through the study of hymn books published between 1913 and 1978. Statistical analysis presented.

GERMAN REFORMED

311. Westermeyer, Paul. "What Shall We Sing in a Foreign Land?: Theology and Cultic Song in the German Reformed and Lutheran Churches of Pennsylvania, 1830–1900." Ph.D. dissertation. Illinois: University of Chicago, 1978. vi, 270 p.

Examines music and liturgy of the German Reformed and Lutheran churches of nineteenth-century Pennsylvania. Bibliography of approximately three hundred writings and 120 Service books, hymnals, and collections of music.

HARMONISTS

312. Wetzel, Richard D. *Frontier Musicians on the Connoquenessing, Wabash, and Ohio: A History of the Music and Musicians of George Rapp's Harmony Society (1805–1906).* Athens: Ohio University Press, 1976. xi, 294 p. ISBN 0–8214–0208–0 ML200.4 .W48

Account of the economic, religious, and social aspects of the nineteenth-century German-American communal society, George Rapp's Harmony

Society (known as the Harmonists). Includes discussion of their music, vocal and instrumental ensembles, and performance practices. Numerous musical examples, illustrations, photographs, and facsimiles; classified bibliography of the music collection of Economy Village, Ambridge, Pennsylvania (more than five hundred printed instrumental and vocal scores and parts, hymns, psalters, and theory and pedagogical books; more than 250 manuscript instrumental and vocal scores and parts, hymns and hymnals, and theory and pedagogical books); general bibliography of nearly seventy writings; index. Accompanied by a recording of representative music.

HISPANIC AMERICAN

313. Lockwood, George F. "Recent Developments in U. S. Hispanic and Latin American Protestant Church Music." D.Min. dissertation. California: School of Theology at Claremont, 1981. ix, 165 p.

Purpose: "to chronicle and analyze the growth of the movement to create new and indigenous music for the Protestant churches in Latin America and the Hispanic churches in the United States of the last two decades (1960–1980)." Musical examples and tables; bibliography of seventeen writings.

HOLINESS

314. Dargan, William T. "Congregational Gospel Songs in a Black Holiness Church: A Musical and Textual Analysis." Ph.D. dissertation. Middletown, Conn.: Wesleyan University, 1982. xxi, 397 p.

Musical practices in New Born Church of God and True Holiness, New Britain, Connecticut, considered by the author as a representative African-American holiness church. Transcriptions of 104 songs with analysis of textual and melodic form. Investigates the significance of songs sung in the church as musical and ritual forms and as a religious experience; examines the interrelationship of these songs and the three techniques of interpreting gospel. Musical examples and charts; bibliography of about 120 writings.

INDEPENDENT CHRISTIAN

315. Ihm, Dana Elizabeth. "Current Music Practices of the Independent Christian Churches in the United States." Ph.D dissertation. Columbia: University of South Carolina, 1994. xiv, 198 p.

Describes current music practices in the Independent Christian Churches in the United States. 535 congregations surveyed. Introduces topic with brief history of Independent Christian Church music practices. Numerous tables; bibliography of forty-three writings.

JEWISH

316. Glazerson, Matityahu. *Music and Kabbalah.* Rev. by R. Harris. Jerusalem: Yerid ha-sefarim, 1988. 130, 17 p.; reprint, Northvale, N.J.: Jason Aronson, 1997. xii, 109 p. ISBN 1–5682–1933–4 (U.S.) ML3195 .G5313, ML3195 .G5313 1997

    Musical practices of the Jewish Synagogue. Addresses musical scales, terminology, and instruments. General study; not specifically related to music in the United States. No bibliography; expansive index. The 1988 edition includes transcriptions of Shabbat and festival songs composed by Glazerson; the 1997 edition does not include the transcriptions.

317. Hart, Stephen A. "An Historical Overview of the Function of Music in the American Reform Synagogue." Ordination thesis. Cincinnati, Ohio: Hebrew Union College-Jewish Institute of Religion, 1986. iii, 143 p.

    Examines musical traditions of the American Reform Synagogue. One musical facsimile; classified bibliography of eighty-three writings.

318. Heskes, Irene. *Passport to Jewish Music: Its History, Traditions, and Culture.* Westport, Conn.: Greenwood Press, 1994. xii, 353 p. ISBN 0–313–28035–5 ML3776 .H47 1994; reprint, New York: Tara, 1995. ISBN 0–933676–45-X ML3776 .H47 1995

    One section, comprising fifty pages, is devoted to "Three Hundred Years of Jewish Music in America." Covers sacred and secular music. Selected topics include hymnology, cantorial artistry, schools for cantorial study, the hymn *Eili, Eili,* and the *Reform Union Hymnal.* Documented with endnotes; index.

319. Hieronymus, Bess. "Organ Music in the Worship Service of American Synagogues in the Twentieth Century." D.M.A. dissertation. Austin: University of Texas, 1969. vii, 107 p.

    Purpose: "to present . . . an understanding of the role of music, particularly the organ in American Temples, and to show how the historical events in the life of the Jewish people have prevented early inclusion of the instrument in the worship service." Chapter 6, "Biographies of Composers of Organ Music," includes biographies of a dozen European-born composers who made their careers in the United States and a few

American-born composers. Musical examples; bibliography of thirty-five writings.

320. Isaacs, Ronald H. *Jewish Music: Its History, People, and Song.* Northvale, N.J.: Jason Aronson, 1997. xiv, 247 p. ISBN 0–7657–5966–7 ML3195 .I83 1997

Not specifically on Jewish music in the United States, but several sections address the subject. Chapter 8 discusses the role of the cantor and lists names with biographical information of approximately 150 cantors, several from the United States. Likewise, Chapter 15 lists approximately 150 twentieth-century Jewish musicologists, composers, and musicians, many from the United States. Chapter 13 provides a brief discussion of African-American spirituals. Chapter 14 covers Jewish American rock music. Other U.S. topics throughout. Musical examples and illustrations; bibliography of nine writings; expansive index.

321. Koskoff, Ellen. *Music in Lubavitcher Life.* Urbana: University of Illinois Press, 2001. xviii, 225 p. ISBN 0–252–02591–1 ML3195 .K65 2001

Study of the musical practices of a community of Lubavitcher Hasidim in Crown Heights, Brooklyn, New York. Musical examples and photos; glossary of terms; discography of twenty-nine recordings; bibliography of nearly 140 writings, including several interviews; expansive index.

322. Levine, Joseph A. *Synagogue Song in America.* Crown Point, Ind.: White Cliffs Media Company, 1989. xxii, 232 p. + 3 audio cassettes. ISBN 0–941677–12–5, ISBN 0–941677–14–1 (pbk.) ML3195 .L48 1989; reprint, Northvale, N.J.: Jason Aronson, 2000. ISBN 0–7657–6139–4 ML3195 .L48 2000

Examines psalmodic, biblical, modal, and performance techniques of Synagogue song, with specific mention of common practices in the United States. Musical examples, photographs, and charts; glossary of terms; useful appendixes on Lithuanian biblical chant and *Misinai* tunes; no bibliography; expansive index. Accompanied by three one-hour audio cassettes of representative musical selections.

323. Slobin, Mark. *Chosen Voices: The Story of the American Cantorate.* Urbana: University of Illinois Press, 1989. xxv, 318 p. ISBN 0–252–01565–7 ML3195 .S55 1989

Historical review of the American cantorate from the 1680s to the 1980s. Includes interviews with cantors. Addresses the role of the cantorate in the Synagogue and examines the music currently performed. Musical examples and photographs; bibliography of approximately 230 writings; expansive index.

324. Summit, Jeffrey A. *The Lord's Song in a Strange Land: Music and Identity in Contemporary Jewish Worship.* Oxford: Oxford University Press, 2000. xiii, 203 p. + 1 compact disc. ISBN 0–19–511677–1 ML3195 .S96 2000

Examines music practices of five diverse Jewish worship communities in Boston, Massachusetts. Accompanied by a compact disc recording of representative music. Musical examples and photographs; glossary of Hebrew and Yiddish terms; bibliography of nearly 180 writings; expansive index.

325. Werner, Eric. *In the Choir Loft: A Manual for Organists and Choir Directors in American Synagogues.* New York: Union of American Hebrew Congregations, 1957. 54 p. ML3195 .W4

Briefly surveys the development of American synagogue music. Examines tonality, stylistic characteristics, and form and structure of the music. Includes a classified bibliography of then-contemporary compositions and hymn and song collections for the American synagogue; lists composer, title, and an indication of easy, medium, or difficult performance levels. Concludes with suggestions for successful programming. Numerous musical examples; no index.

KOREAN AMERICAN

326. Kim, Eun Chul. "A Study of Current Practice of Liturgical Tradition in Selected Korean-American Churches with a Special Emphasis on Music." Thesis of independent study. Princeton, N.J.: Princeton Theological Seminary, 1992. v, 71 p.

Identifies differences in worship style and music selection of Korean-speaking and English-speaking Korean-American churches in the United States. Advocates developing a worship format to establish mutuality between the two. Research based on interviews and surveys. Graphs and charts; bibliography of nearly fifty writings and a couple of hymnals.

LUTHERAN

327. Blume, Friedrich, in collaboration with Ludwig Finscher, Georg Feder, Adam Adrio, Walter Blankenburg, Torben Schousboe, Robert Stevenson, and Watkins Shaw. *Protestant Church Music: A History.* Foreword by Paul Henry Lang. New York: W. W. Norton, 1974. xv, 831 p. ISBN 0–393–02176–9 ML3100.B5913; reprint, London: Gollancz, 1975. ISBN 0–575–01996–4 ML3100 .B5913

English translation of *Geschichte der evangelischen Kirchenmusik* (1965) with revisions and additional sections and chapters. Nine sections by various authors deal with the history of the Lutheran tradition. One fifty-four-page section entitled "Protestant Music in America" by Robert Murrell Stevenson addresses church music in the United States from 1564 to the 1960s. Numerous musical examples, illustrations, and facsimiles; classified bibliography of about fifteen hundred entries (general history, church history, and theology; music history and church history; chorales and hymn books; historia and passion; organ music; and monographs on individuals); expansive index.

328. Brown, Bruce Calvin. "The Choral Music of the American Lutheran Church: A Study and Performance of the Music of Daniel Moe, Paul Fetler, and Rolf Espeseth." D.M.A. dissertation. Los Angeles: University of Southern California, 1974. iv, 223 p.

History of choral composition in the Lutheran Church focusing on the influence of the F. Melius Christiansen choral tradition and the compositional output of Moe, Fetler, and Espeseth. Several appendixes, including catalogs of the choral music of Moe, Fetler, and Espeseth. Concludes with editions of five works by Espeseth. Musical examples; bibliography of twenty-nine writings and interviews.

329. Liemohn, Edwin. *The Chorale through Four Hundred Years of Musical Development as a Congregational Hymn.* Philadelphia: Muhlenberg Press, 1953. xii, 170 p. ML3184 .L5

Study begins in the early sixteenth century and progresses chronologically to the twentieth century, concentrating on the development of the chorale in Germany, Sweden, Norway, and Denmark. Sixth chapter discusses the chorale in America with emphasis on churches of German, Swedish, Norwegian, and Danish backgrounds. Musical practices of the Lutheran church is the main focus of this section. Musical examples; bibliography of fifty writings and 122 music collections and hymnals.

330. Schalk, Carl. *God's Song in a New Land: Lutheran Hymnals in America.* St. Louis, Mo.: Concordia, 1995. 239 p. ISBN 0–570–04830–3 BV410.A1 S33 1995

Described by the author as "an introduction to the history" of Lutheran hymnals in the United States from 1786 to the present. Addresses German and Scandinavian influences. Facsimiles, illustrations, and tables; documented with endnotes; index.

331. Stulken, Marilyn Kay. *Hymnal Companion to the Lutheran Book of Worship.* Introduction by Eugene L. Brand. Philadelphia: Fortress Press, 1981. xxiii, 647 p. ISBN 0–8006–0300–1 ML3168 .S85

Organized into two parts: (1) historical essays and (2) the canticles and hymns. Among the historical essays is "English and American (Non-Lutheran) Hymnody" (S. E. Yoder). The portion of the essay that discusses American hymnody is only a few pages, covering *The Bay Psalm Book* (1640) through the nineteenth century. The historical essay is followed by a thirty-three-page essay titled "Lutheran Hymnody in North America" (R. H. Terry) which primarily surveys significant Lutheran hymnals and various Lutheran ethnic groups. The second part of the book discusses each of the 569 canticles and hymns published in the *Lutheran Book of Worship* (Augsburg Publishing House; Board of Publication, Lutheran Church in America, 1978). Musical examples; bibliography of twenty-four writings. Several indexes: index of hymns by occasion; index of tunes classified by country of origin and listed in chronological order; index of original-language first lines of hymns classified by language of origin and listed in chronological order; index of authors, composers, and sources of hymns and canticles; alphabetical index of titles; alphabetical index of first lines of canticles; alphabetical index of first lines of hymns, original and translated; index to essays.

332. Wolf, Edward C. "Lutheran Church Music in America during the Eighteenth and Early Nineteenth Centuries." Ph.D. dissertation. University of Illinois at Urbana-Champaign, 1960. xviii, 454 p.

Three aims: (1) to record and evaluate information regarding the position music held in Lutheran life in America before approximately 1825, (2) to assess the relation of Lutheran musical activities to the history of American church music in general, and (3) to provide examples which would be applicable to church music today. Numerous appendixes, which include musical texts, excerpts of writings, and music. Musical examples; bibliography of approximately 185 writings and music.

See also: Ode, James A. *Brass Instruments in Church Services* (item 470); Westermeyer, Paul. "What Shall We Sing in a Foreign Land?: Theology and Cultic Song in the German Reformed and Lutheran Churches of Pennsylvania, 1830–1900" (item 311)

MENNONITE

333. Faus, Nancy Rosenberger, ed. *The Importance of Music in Worship.* Elgin, Ill.: Brethren Press; Newton, Kan.: Faith and Life Press; Scottdale, Pa.: Mennonite Publishing House, 1993. 45 p. BV403.C16 I47 1993

A collection of three essays: "Music in the Mennonite Church" (P. K. Clemens); "Music in the Church of the Brethren" (Rosenberger); and

"Hymnody of the General Conference Mennonite Church" (M. Houser Hamm). Documented with endnotes. Message presented: "Mennonites and Brethren need to acknowledge that singing plays different roles in the three denominations as well as in the various congregations within each church body."

334. Klassen, Roy Leon. "The Influences of Mennonite College Choral Curricula upon Music Practices in American Mennonite Churches." D.M.A. dissertation. Tempe: Arizona State University, 1990. ix, 183 p.

Musical practices of selected Mennonite churches in comparison to choral music curricula of selected U.S. Mennonite colleges. Numerous charts; bibliography of fifteen writings.

335. Kropf, Marlene, and Kenneth James Nafziger. *Singing: A Mennonite Voice.* Foreword by John L. Bell. Scottdale, Pa.: Herald Press, 2001. 175 p. ISBN 0–8361–9152–8 ML3169 .K76 2001

Asks the question from a theological perspective: "What happens when you sing?" Research based on interviews with more than one hundred people of Mennonite faith in the United States and Canada. Musical examples; documented with endnotes; index of hymn titles.

MENNONITE BRETHREN

336. Friesen, Dietrich. *The Development of Church Music in the Mennonite Brethren Churches.* Introduction by Larry Warkentin. Fresno, Calif.: Fresno Pacific College, 1983. 45 p.

Examines hymn singing, choral singing, and worship service music of the Mennonite Brethren Church in Russia, South America, and the United States. Musical examples.

METHODIST AND UNITED METHODIST

337. Kindley, Carolyn E. "*Miriam's Timbrel:* A Reflection of the Music of Wesleyan Methodism in America, 1843–1899." D.A. dissertation. Muncie, Ind.: Ball State University, 1985. v, 329 p.

Examines nineteenth-century musical practices of Wesleyan Methodism in the United States. Presents analysis of forty-four tunes from *Miriam's Timbrel* (1865), a hymnal published for the Methodist Connection. Musical examples, facsimiles, and charts; bibliography of nearly eighty writings and forty hymnals.

338. McClain, William B. *Come Sunday: The Liturgy of Zion.* Foreword by Woodie W. White. Nashville, Tenn.: Abingdon Press, 1990. 175 p. ISBN 0–687–08884–4 BR563.N4 M33 1990

Liturgical and music practices of African-American congregations of the United Methodist Church in the United States. Companion to the songbook *Songs of Zion* (Abingdon Press, 1981). Second half devoted to the survey of various types of songs in *Songs of Zion,* including spirituals, hymns, gospel songs, and songs for special occasions. Appended: example of a liturgy; discography of nearly forty recordings; bibliography of approximately fifty writings; scripture index and topical index to *Songs of Zion.*

339. Rice, William C. "A Century of Methodist Music: 1850–1950." Ph.D. dissertation. Iowa City: University of Iowa, 1953. vi, 471 p.

History of Methodist church music from 1850 to 1950. Examines significant hymnals, musical practices, and music in Methodist seminaries. Tables; bibliography of 840 writings, thirty-seven hymn books, and a few letters.

340. Volland, Linda L. "Three Centuries of Methodist Hymnody: An Historical Overview of the Development of the American Methodist Hymnal with Special Attention to Hymnody in the 1780, 1878, and 1989 Hymnals." D.M.A. dissertation. Lincoln: University of Nebraska, 1995. i, 385 p.

Studies hymnals and hymnody of the United Methodist Church in the United States, eighteenth through twentieth centuries. Focus on three Methodist hymnals: *A Collection of Hymns for the Use of the People Called Methodists* (1780), *Hymnal of the Methodist Episcopal Church with Tunes* (1878), and *The United Methodist Hymnal* (1989). Various appendixes, which accounts for at least 75 percent of the text, include lists of tunes commonly set to music and lists of tunes found in the three hymnals cited above with commentary on composers, librettists, and stylistic characteristics. Facsimiles; bibliography of approximately fifty writings and thirty hymnals.

341. Warren, James I. *O for a Thousand Tongues: The History, Nature, and Influence of Music in the Methodist Tradition.* Grand Rapids, Mich.: Francis Ashbury Press, 1988. 303 p. ISBN 0–310–51530–0 ML3170 .W34 1988

Broad study of Methodist church music in the United States. Includes hymnody, gospel hymns, and protest songs of the 1960s. Documented with end notes; index of song titles, first lines, and quotations, and expansive indexes of persons and subjects.

See also: Clark, Linda J. *Music in Churches: Nourishing Your Congregation's Musical Life* (item 302)

METHODIST EPISCOPAL

342. Baldridge, Terry L. "Evolving Tastes in Hymntunes of the Methodist Episcopal Church in the Nineteenth Century." Ph.D. dissertation. Lawrence: University of Kansas, 1982. xv, 477 p.

Introduced by a survey of main musical trends in nineteenth-century American hymnody, including shape-note hymnody, folk hymnody, revival hymnody, gospel hymnody, and the contributions of Lowell Mason and the "Better-Music" movement. Examines published music of the Methodist Episcopal Church. Focuses upon evolving musical tastes expressed in writings by Methodists. Compares hymntunes in the denomination's tunebooks with hymnals to identify alterations that may demonstrate evolving tastes. Musical examples and facsimiles; index of tune names found in the 1808, 1822, 1833, 1837, 1848, 1849, 1857, 1866, and 1878 Methodist Episcopal hymnals; bibliography of about 260 writings.

343. Graham, Fred Kimball. *"With One Heart and One Voice": A Core Repertory of Hymn Tunes Published for Use in the Methodist Episcopal Church in the United States, 1808–1878.* Lanham, Md.: Scarecrow Press, 2004. xvi, 167 p. ISBN 0–8108–4983–6 ML3170 .G683 2004

Originated as the author's dissertation by the same name (Drew University, 1991). Study of Methodist Episcopal church music "officially sanctioned for public worship" during the first century of the denomination in the United States. Prefaced by a survey of the state of hymn singing in America during colonial times through the eighteenth century. Chapter 4 is an annotated bibliography of the core repertory of seventy-six hymn tunes published in tunebooks between 1808 and 1878. Provides composer, title, first publication or first U.S. publication, original text if relevant, description of music and characteristics of tune or voicing, comparison to other settings examined in the study, and text incipit. Several helpful appendixes, including an annotated bibliography of tunebooks used in the survey, an alphabetical list of tunes, a metrical index of tunes, an alphabetical list of composers, and a chronological list of tune appearances. Musical examples and facsimiles; bibliography of fifty-four writings.

MORAVIAN

344. Adams, Charles B. *Our Moravian Hymn Heritage: Chronological Listing of Hymns and Tunes of Moravian Origin in the American Moravian Hymnal of 1969.* Bethlehem, Pa.: Department of Publications, Moravian Church in America, 1984. vi, 144 p. ML3172 .A3 1984

Handbook for the American *Moravian Hymnal* of 1969 containing 219 hymns and eighty tunes written/composed by members of the Moravian

Church between 1415 and 1964. There are sixteen hymns and fourteen tunes from the American Provinces, 1871 and 1964. Musical incipits along with historical information about the authors, composers, and music. Indexes provided for (1) authors and translators of hymns, (2) composers, arrangers, and sources of tunes, and (3) first lines of hymns; cross references to hymnal; bibliography of nine writings.

345. Duncan, Timothy Paul. "The Role of the Organ in Moravian Sacred Music between 1740–1840." D.M.A. dissertation. University of North Carolina at Greensboro, 1989. ix, 152 p.

Organized into five parts: (1) Moravian history and music; (2) religious services and music genres that incorporate organ music; (3) the role of the organ in regard to hymn tunes, hymn texts, active chorale repertoire, and hymn accompaniment; (4) the duties of the organist; and (5) organ design. Musical examples and photos; bibliography of seventy-six writings.

346. Fox, Pauline Marie. "Reflections on Moravian Music: A Study of Two Collections of Manuscript Books in Pennsylvania ca. 1800." Ph.D. dissertation. New York: New York University, 1997. xiv, 272 p.

An historical introduction traces the development of Moravian church in Europe, its migration to America, and selected Moravian boarding schools in Pennsylvania. Surveys the contents of two collections of Moravian music known as the Bethlehem Manuscript Books (ca. 1760–1837) and the Lititz Manuscript Books (ca. 1800–1846), along with biographical sketches of previous owners of the books. Bibliography of more than one hundred writings; expansive index.

347. Gombosi, Marilyn P., ed. *Catalog of the Johannes Herbst Collection.* Chapel Hill: University of North Carolina Press, 1970. xix, 255 p. ISBN 0–8078–1124–6 ML136.C44 U64 1970

Catalog of the Johannes Herbst Collection held at The Moravian Archives in Winston-Salem, North Carolina. The collection was once part of the private library of Johannes Herbst (1735–1812), a Moravian minister and musician. Provides composer, title, musical incipit, and description of music. Indexes for composers and titles.

348. Hall, Harry H. "The Moravian Wind Ensemble: Distinctive Chapter in America's Music." Ph.D. dissertation. Nashville, Tenn.: Peabody College at Vanderbilt University, 1967. 2 vols.: xiii, 418 p.; iv, 133 p.

Two volumes: (1) historical survey and (2) musical supplement. Vol. 1: Surveys the musical traditions, specifically the role of the wind ensemble, of the Moravian church in the United States from its beginning in Georgia in the eighteenth century through the 1960s, with primarily focus on musical traditions in Pennsylvania and North Carolina. Vol. 2: Includes twenty compositions performed primarily by Moravian wind groups of the late eighteenth and early nineteenth centuries. Bibliography of approximately 250 writings and thirty music publications.

349. Hartzell, Lawrence W. *Ohio Moravian Music.* Winston-Salem, N.C.: Moravian Music Foundation Press; Cranbury, N.J.: Associated University Presses, 1988. 201 p. ISBN 0–941642–02-X ML3172 .H37 1988

Historical survey of the musical practices of Indian missions and white settlements of the Moravian Church in Ohio over a two-hundred-year period (1761–1961). Draws information from diaries, church records, interviews, and other sources. Provides brief biographies of a number of significant musicians. Facsimiles; lists with dates of service of known ministers, organists, and choir, youth choir, and instrumental music directors; bibliography of nearly 250 writings; expansive index.

350. Knouse, Nola Reed, and C. Daniel Crews. *Moravian Music: An Introduction.* Winston-Salem, N.C.: Moravian Music Foundation, 1996. 33 p. ML3172 .K58 1996

Introductory study of Moravian music in the United States in the late eighteenth and early nineteenth centuries. Biographical sketch of ten notable composers; facsimiles and photos; glossary of a few music terms; bibliography of forty-eight writings.

351. Knouse, Nola Reed. *Opening a Can of Worms: Reflections on Music and Worship in Today's Moravian Church.* Winston-Salem, N.C.: Moravian Music Foundation, 1996. 33 p.

Essay on the role of music in Moravian church worship. The "can of worms" is expressed in following statement attributed to Donald Hustad: "None of the world's music is incapable of fruitful use in expressing the gospel of Jesus Christ." Bibliography of seventeen writings.

352. Kroeger, Karl. "The Moravian Choral Tradition: Yesterday and Today." *The Choral Journal,* 19/5 (Jan. 1979): 5–9, 12

Kroeger, then-Director of The Moravian Music Foundation, reviews past and present choral music and performance practices of the Moravian church through study of the Moravian archives. Provides a selected list of twenty-six modern editions of Moravian choral works and collections prepared by The Moravian Music Foundation. Photographs and one facsimile of music; documented with footnotes.

353. McCorkle, Donald. "Moravian Music in Salem: A German-American Heritage." Ph.D. dissertation. Bloomington: Indiana University, 1958. ix, 414 p.

Surveys musical culture of the Moravians in Salem, Massachusetts, between 1780 and 1840. Examines chorales, hymns, anthems, arias, and role of the organ. Several useful appendixes, including composer/title index to Johannes Herbst Collection of sacred vocal and choral music, composer/title index to Salem Congregation Collection of sacred vocal and choral music, and annotated, classified bibliography of manuscripts before 1825 in the Johann Friedrich Peter Collection. Musical examples and photos; classified bibliography of nearly 250 writings; discography of a few recordings.

354. Runner, David Clark. "Music in the Moravian Community of Lititz." D.M.A. dissertation. New York: University of Rochester, 1976. vii, 90 p.

Sacred and secular musical activities in early Lititz, Pennsylvania, during the eighteenth century and through the first half of the nineteenth century. Appended are lists of the Lititz Collegium Musicum collection, music at Linden Hall Seminary, and musical instruments in the Lititz church museum. Bibliography of approximately eighty writings.

355. Sharp, Timothy W. "Moravian Choral Music." *The Choral Journal,* 30/3 (Oct. 1989): 5–7, 9–12

Choral music of the Moravian Church in the United States from the mid-eighteenth century to the 1980s. Discusses significant publications of hymns, anthems, cantatas, etc., and the pioneering work of The Moravian Music Foundation. Includes a classified listing (SATB, SSAB, SSTT, SAB, two parts, multiple choirs, solos, and collections) of approximately 165 published Moravian choral works. Documented with endnotes.

See also: Grider, Rufus A. *Historical Notes on Music in Bethlehem, Pennsylvania from 1741 to 1871* (item 196); Larson, Paul. *An American Musical Dynasty: A Biography of the Wolle Family of Bethlehem, Pennsylvania* (item 39)

MORAVIAN BRETHREN

356. Lawson, Charles T. "Musical Life in the Unitas Fratrum Mission at Springplace, Georgia, 1800–1836." Ph.D. dissertation. Tallahassee: Florida State University, 1970. v, 174 p.

Study of use of music at the Moravian Brethren mission established in Springplace, Georgia, in the early nineteenth century for the purpose of teaching Cherokee Indians. Illustrations, photos, and facsimiles; bibliography of seventy writings.

MORMON

357. Bentley, Brian Richard. "The Philosophical Foundations and Practical Use of Choral Music in the Church of Jesus Christ of Latter-Day Saints in the Twentieth Century, Focusing on the Music of Robert Cundick, with Emphasis on His Sacred Service, *The Redeemer*." D.M.A. dissertation. Ohio: University of Cincinnati, 1999. xiii, 312 p.

    Studies the use of choral music and some vocal music in Mormon worship services, with emphasis on the prominent composer, Robert Cundick, and his oratorio, *The Redeemer: A Sacred Service.* Covers the Mormon Tabernacle Choir, hymns, music in church-sponsored public education, music education within the infrastructure of the church, church leaders' opinions and teachings on music, and "Mormon" music. General bibliography of approximately one hundred writings, twenty Web sites, and a few scores and recordings; mammoth bibliography of choral music by selected twentieth-century Mormon composers, approximately 110 composers and three thousand titles, arranged in two listings: (1) alphabetical by composer/arranger and (2) alphabetical by title. Provides composer/arranger, title, accompaniment, voicing, and publisher information.

358. Hannum, Harold Byron. *Music and Worship.* Nashville, Tenn.: Southern Publishing Association, 1969. 128 p. ML3000.H162 M9

    Described by the author as "a philosophy of church music." Views relate to the Mormon Church. Seventeen chapters discuss a variety of topics, including: aesthetics and religion; music and evangelism; sacred and secular; music and education; twelve great hymns; mediocrity in sacred music; choral music and solos; and worship and instrumental music. Bibliography of fifty-three writings; expansive index.

359. Hicks, Michael. *Mormonism and Music: A History.* Urbana: University of Illinois Press, 1989. xii, 243 p. ISBN 0–252–01618–1 ML3174 .H5 1989

    Survey of music within the Mormon Church from the 1830s through the 1970s. Discusses hymnology, role of choral music, and use of instruments in church. Entire chapter dedicated to the history of the Mormon Tabernacle Choir. Musical examples and photos; documented with endnotes; expansive general index and index of first lines and titles of hymns and songs.

360. Moody, Michael Finlinson. "Contemporary Hymnody in the Church of Jesus Christ of Latter-Day Saints." D.M.A. dissertation. Los Angeles: University of Southern California, 1972. iv, 163 p.

    Examines contemporary Mormon hymnody since 1950 with the objective "to motivate poets and musicians in the Church toward an era of increased enthusiasm and productivity." Introduced by a history of Mormon hymnody from 1830 to 1950. Musical examples; several appendixes, including a bibliography of more than three hundred hymns by nearly 140 hymn authors; bibliography of approximately 115 writings and thirteen hymnals.

361. Wolford, Darwin. *Song of the Righteous.* Foreword by Robert R. Worrell. Springville, Utah: CFI, 1995. xv, 355 p. ISBN 1–55517–193–1 BX8643.M8 W6 1995

Discusses the lives of fifteen leaders in the Church of Jesus Christ of Latter-Day Saints, including Joseph Smith and Brigham Young. Focuses on music in the home and in worship. Illustrations; documented with endnotes; expansive index.

NATIVE AMERICAN

362. Davidson, Jill D. "Prayer-Songs to Our Elder Brother: Native American Church Songs of the Otoe-Missouria and Ioway." Ph.D. dissertation. University of Missouri-Columbia, 1997. 2 vols.: vii, 546 p.

Two volumes: Vol. 1 is a study of the Native American Church songs in the language of the Otoe-Missouria and Ioway commonly known as Chiwere (Siouan). Vol. 2 is a glossary of English Peyote terms, a 291-page bibliography of Native American Church song texts with interlinear notes and free translations along with an index of composers and songs, a ten-page bibliography of other Otoe-Missouria and Ioway song texts organized by genre (patriotic songs; Eroska song; church hymn), and four historical and ethnographic documents. Bibliography of approximately 120 writings.

See also: Ray, Mary Dominic, and Joseph H. Engbeck. *Gloria Dei: The Story of California Mission Music* (item 291)

PENTECOSTAL

363. Alford, Delton L. *Music in the Pentecostal Church.* Foreword by Donald S. Aultman. Cleveland, Tenn.: Pathway Press, 1967. 120 p. ML3178.P4 A4

Organized into two parts. Part I describes representative Pentecostal church music in context of significant religious and historical influences. Part II addresses the music through worship, evangelism, and Christian education. Classified bibliography of thirty-four choral works for graded choirs; bibliography of twenty-five writings.

364. Guthrie, Joseph R. "Pentecostal Hymnody: Historical, Theological, and Musical Influences." D.M.A. dissertation. Fort Worth, Tex.: Southwestern Baptist Theological Seminary, 1992. xi, 187 p.

Study of Pentecostal hymnody with emphasis on songs of the holiness movement, camp meeting song, and gospel song. Notes significant influences on Pentecostal church music, including Wesleyan hymnody and charismatic style of worship. Musical examples and tables; bibliography of seventy-three writings and twelve hymnals.

See also: Neely, Thomasina. "Belief, Ritual, and Performance in a Black Pentecostal Church: The Musical Heritage of the Church of God in Christ" (item 301)

PRESBYTERIAN

365. Busche, Henry E. *Presbyterian Church Music: What is Its Legacy?*. S.1.: Busche, 1999. ii, 114, 53 p. ML3176 .B87 1998

A historical study, described as a "sketchy outline" by the author, of music in the Presbyterian Church. Discusses music practices of John (Jean) Calvin and the Reformed Church and the Presbyterian Church in Scotland and the United States. Musical examples; bibliography of nearly fifty writings.

366. Leaver, Robin A. "The Hymnbook as a Book of Practical Theology" in *Exploring Presbyterian Worship: Contributions from Reformed Liturgy & Music*. Introduction by Joseph D. Small. Louisville, Ky.: Distribution Management Services, 1994. p. 33–38. BX9185 .E9 1994

Leaver, Robin A. "The Hymnbook as a Book of Practical Theology." *Reformed Liturgy & Music*, 24/2 (Spr. 1990): 55–57

Exploring Presbyterian Worship: Contributions from Reformed Liturgy & Music is a collection of seven articles previously published in the journal *Reformed Liturgy & Music*. One article directly relates to Presbyterian worship music: "The Hymnbook as a Book of Practical Theology" (Leaver). Leaver describes the hymnbook as a volume of practical theology dealing with the content (content of the hymnbooks), structure (sequence of main headings under which hymns appear), and context of the faith (found in the "Aids to Worship," which precede the hymns).

367. Lester, Joan Stadelman. "Music in Cumberland Presbyterian Churches in East Texas Presbytery, 1900–1977, as Recorded in Church Records and as Related in Oral and Written Interviews." D.M.A. dissertation. Austin: University of Texas, 1981. xv, 264 p.

Music in Cumberland Presbyterian churches from 1900 until they died out around 1977. Examines hymnbooks used and offers descriptions of twentieth-century singing schools. Tables, illustrations, and photographs;

bibliography of nineteen hymnals, about fifty writings, and many interviews and church record documents.

PRIMITIVE BAPTIST

368. Drummond, R. Paul. *A Portion for the Singers: A History of Music among Primitive Baptists since 1800.* Commentary by V. J. Lowrance. Foreword by Daniel W. Patterson. Atwood, Tenn.: Christian Baptist Library & Publishing Company, 1989. xxx, 486 p. ML3160 .D7 1989

 Originated as the author's doctoral dissertation, "A History of Music among Primitive Baptists since 1800" (University of Northern Colorado, 1986). Study of "Old Baptist" singing among Primitive Baptists of the southern and southwestern regions of the United States. Examines four musical traditions: (1) southern folk, (2) Mason/Bradbury hymns, (3) gospel songs, and (4) traditional Protestant hymns. List of approximately 130 choral arrangements of American folk hymns, providing title, arranger, tune name, octavo number, publisher, and voicing with accompaniment. Musical examples, facsimiles, illustrations, photographs, and tables; bibliography of around three hundred writings and seventy hymnals and tuneless hymnbooks. No index; however, the table of contents is an outline of the study.

369. Patterson, Beverly Bush. *The Sound of the Dove: Singing in Appalachian Primitive Baptist Churches.* Urbana: University of Illinois Press, 1995. x, 238 p. + 1 audio cassette. ISBN 0–252–02123–1 (text), ISBN 0–252–02173–8 (cassette) ML3160 .P28 1995

 Based on fieldwork among Primitive Baptists (North Carolina, Virginia, West Virginia, and eastern Kentucky), whose music is characterized by unaccompanied singing. Originated from the author's doctoral dissertation (University of North Carolina at Chapel Hill, 1989). Provides historical and textual information about the music, insight into performance practice, and addresses social and cultural patterns among church members. Musical examples, facsimiles, and photos; bibliography of nearly 120 writings; expansive index.

PURITAN

370. Campbell, Donald Perry. "Puritan Belief and Musical Practices in the Sixteenth, Seventeenth, and Eighteenth Centuries." D.M.A. dissertation. Fort Worth, Tex.: Southwestern Baptist Theological Seminary, 1994. iv, 558 p.

 Rather extensive study of Puritan worship practices in England and the United States. As a comparison, also examines traditional Episcopal/Anglican ("non-Puritan") church music practices. Bibliography of more than two hundred writings.

371. Scholes, Percy Alfred. *The Puritans and Music in England and New England: A Contribution to the Cultural History of Two Nations.* London: Oxford University Press/Humphrey Milford, 1934; New York: Russell & Russell, 1934. xxii, 428 p. ML194.S3 M9; reprint, New York: Russell & Russell, 1962. ML194.S3 M9 1962; reprint, Oxford: Clarendon Press, 1969. ML194.S3 M9 1969

Primarily devoted to Puritans and music in England. A few latter chapters discuss music in the American colonies in detail: Chapter 16, "Puritan Church Song in England and New England," Chapter 17, "What the Psalms Meant to the Puritans," Chapter 19, "A Glimpse at Musical Life in a Non-Puritan Colony," and Chapter 21, "The Eighteenth-Century Origin of the Objection to Music." Illustrations, photos, and facsimiles; glossary of terms; index of works cited and classified, expansive index (with separate index section on "America").

See also: Clark, Linda J. *The Sound of Psalmody* (item 478); Haraszti, Zoltán. *The Enigma of the Bay Psalm Book* (item 482)

SEVENTH-DAY ADVENTIST

372. Pierce, Charles L. "A History of Music and of Music Education of the Seventh-Day Adventist Church." D.M.A. dissertation. Washington, D.C.: Catholic University of America, 1976. vi, 287 p.

Two parts: (1) Music of the Seventh-Day Adventist Church from 1849 to 1975 and (2) history of music education in Adventist colleges. Addresses church music, namely gospel song, the hymn, choral music, sacred instrumental compositions, and oratorio and cantata by approximately sixty composers. Five appendixes: (1) hymns in Adventist hymnals by Adventist composers; (2) photocopied examples of some of the composers' hymns; (3) song books and hymnals published by the Adventist Church; (4) list of music teachers in Adventist colleges, dating from 1876 to 1965; and (5) guidelines toward an Adventist philosophy of music. Bibliography of fifty-three writings.

SHAKERS

373. Anderson, Cheryl P. *Shaker Music and Dance: Background for Interpretation.* Pittsfield, Mass.: Education Dept., Shaker Community, Inc., 1984. 21, 36 p.

Manual "intended to provide the bare bones of music and dance interpretation at Hancock Shaker Village, [Massachusetts]." Includes fifty-one songs collected and transcribed by the author with assistance from Meg Robertson. Documented with endnotes. Illustrations.

374. Bell, Vicki P. "Shaker Music Theory: The Nineteenth-Century Treatises of Isaac Newton Youngs and Russel Haskell." Ph.D. dissertation. Lexington: University of Kentucky, 1998. x, 267 p.

Examines the nineteenth-century treatises of Youngs and Haskell concerning "small letteral notation," a music notation based on a system of letters used in the music of the Shakers. Focuses on notation, melody, rhythm, and meter.

375. Christenson, Donald Edwin. "Music of the Shakers from Union Village, Ohio: A Repertory Study and Tune Index of the Manuscripts Originating in the 1840's." Ph.D. dissertation. Columbus: Ohio State University, 1988. xii, 608 p.

Topic introduced by two chapters (112 pages) that broadly survey religious music in the United States from the seventeenth century to the Civil War, covering Roman Catholic, Protestant, Anabaptists, Pietists, Jewish faith, Russian and Greek Eastern Orthodox, and communal societies. The third and following chapters specifically discuss music of the Shakers. Provides an inventory of eight hundred extant Shaker manuscripts. Examines eleven manuscripts from that collection in greater detail. Chosen manuscripts all date from the period known as Mother Ann's Work (ca. 1837–1847). Musical examples and facsimiles; bibliography of approximately 120 writings and nearly thirty music manuscripts; several indexes: titles, first lines, authors, and composers; includes a thematic index.

376. Cook, Harold E. *Shaker Music: A Manifestation of American Folk Culture.* Lewisburg, Pa.: Brucknell University Press, 1973. 312 p. ISBN 0–8387–7953–0 ML3178.S5 C6

Originated from the author's doctoral dissertation (Western Reserve University, 1946). Examines the daily life of the Shaker through records, diaries, journals, and letters. Bibliography of approximately 140 writings; collation of approximately six hundred manuscript hymnals from various collections, the largest located at The Western Reserve Historical Society, Cleveland, Ohio; expansive general index and index of songs.

377. Gurd, Franklin Henry. "Shaker Music and Its Influence in America." Project. Fort Worth, Tex.: Southwestern Baptist Theological Seminary, 1976. iv, 169 p.

Project divided into four sections: (1) history of the Shakers; (2) theological perspectives affecting Shaker music; (3) development of Shaker music including performance practice; and (4) influence of Shaker music. Musical examples, illustrations, and photos; bibliography of approximately seventy-five writings.

378. Hall, Roger L. *A Guide to Shaker Music: With Music Supplement.* 4th ed. Stoughton, Mass.: Pinetree Press, 2001. 64 p. ML3178.S5 H26 2001

Potpourri of descriptions, chronologies, list of significant individuals, bibliographies, and a discography relating to Shaker music. Provides the following: a chronology of evolution of Shaker music (1774–1999); list of popular tunes and major tunesmiths and a chronology of Issachar Bates' contribution; bibliography of manuscripts, printed hymnals, and music books containing original Shaker music; bibliography of arrangements for voice and instruments; bibliography of writings which describe Shaker music; excerpts from writings by those who visited Shaker communities during the eighteenth and nineteenth centuries; bibliography of fifty-five writings by and about the Shakers; and discography of thirty-one recordings of original Shaker music and arrangements. Musical supplement consists of fifteen Shaker tunes and texts set by the author.

379. Hall, Roger L. "Shaker Hymnody: An American Communal Tradition." *Journal of Church Music,* 17/8 (Oct. 1975): 2–6

Discusses hymn repertory of the Shakers. Surveys Shaker hymnals. Musical examples.

380. Hall, Roger L. "The Simple Gifts of Shaker Music." *Sing Out!: The Folk Song Magazine,* 43/1 (Sum. 1998): 64–73

Discusses early tunes, hymns, music notation, and recorded music of the Shakers. Musical examples, photos, and illustrations; brief list of recommended recordings and writings.

381. Patterson, Daniel W. *The Shaker Spiritual.* 2nd, corrected ed. Mineola, N.Y.: Dover, 2000. xxv, 562 p. ISBN 0–486–41375–6 ML3178.S5 P4 2000

First published in 1979. Discussion of Shaker spiritual accompanied by over three hundred musical transcriptions. Provides a listing of approximately one thousand Shaker song manuscripts and a few recordings. Photos and illustrations; indexes for first lines and titles, non-Shaker songs cited, and expansive index for persons and subjects.

382. Schaeffer, Vicki. "An Historical Survey of Shaker Hymnody Expressing the Christian Virtues of Innocence and Simplicity." D.M. dissertation. Bloomington: Indiana University, 1992. vii, 70 p.

Brief study of Shaker hymnody with specific focus on hymn texts that relate to the Christian virtues of innocence and simplicity. Annotated list of nine manuscripts; bibliography of thirty writings; index to first line of hymns.

383. Smith, Harold Vaughn. "Oliver C. Hampton and Other Shaker Teacher-Musicians of Ohio and Kentucky." D.A. dissertation. Muncie, Ind.: Ball State University, 1981. vi, 244 p.

Introduces the study with an eight-page history of the Shakers. Subsequent chapters cover Shaker attitudes on religion, education and music, and the contributions of Oliver C. Hampton (1817–1901) and composers Susanna M. Brady, Nancy Rupe, and Polly M. Rupe, among others. Photos and numerous facsimiles; bibliography of more than one hundred writings and music books.

SOUTHERN FUNDAMENTALIST

384. Brown, Marian Tally. "A Resource Manual on the Music of the Southern Fundamentalist Black Church." Ed.D. dissertation. Bloomington: Indiana University, 1974. v, 174 p.

One of the purposes of this writing is to assist teacher preparation in the area of African American studies. Includes outlines and lesson plans. Chapter 4, "The Music and Musical Materials of the Southern Fundamentalist Black Church," reviews the origins and characteristics of the music and its influence upon secular styles within the African-American church tradition. Tables; annotated discography of nearly sixty compositions which demonstrate the influence of traditional African-American church musical styles; bibliography of about sixty-five writings.

SWEDISH AMERICAN

385. Erickson, J. Irving. *Twice-Born Hymns.* Chicago, Ill.: Covenant Press, 1976. 132 p. ML3178.E83 E7 1976

Church music of Swedish Americans. Three parts: (1) historical discussion of the development of hymnody from the Reformation in Sweden to the publication of *The Covenant Hymnal* (Covenant Press, 1973) by the Evangelical Covenant Church of America; (2) study of 104 texts and tunes of the hymns; and (3) biographies of authors, revisers, translators, and composers. Musical examples, illustrations, and photographs; bibliography of nearly sixty writings; index.

UNITED MISSIONARY CHURCH

386. Tweed, Myron Leland. "The Function of Music within the United Missionary Communion." D.M.A. dissertation. Los Angeles: University of Southern California, 1970. xvii, 371 p.

    Examines the historical function of music and current musical practices within the United Missionary Church in the United States and Canada. Proposes a program for the development of music in the United Missionary Church. Tables; bibliography of about 140 writings and eleven hymnals.

VII

Church and Sacred Music Genres

ANTHEMS

387. Dakers, Lionel. *The Church Anthem Handbook: A Companion to the One Hundred Anthems in "The New Church Anthem Book."* Oxford, N.Y.: Oxford University Press, 1994. vii, 67 p. ISBN 0–19–353108–9 M2060 .N54 1992 Suppl.

     Provides general and performance notes for each anthem contained in *The New Church Anthem Book: One Hundred Anthems* (Oxford, N.Y.: Oxford University Press, 1992) compiled and edited by Dakers. Liturgical index.

388. Daniel, Ralph T. *The Anthem in New England before 1800.* With a Foreword by George Howerton. Evanston, Ill.: Northwestern University Press, 1966. xvi, 282 p. ML2911 .D35; reprint, New York: Da Capo Press, 1979. ISBN 0–306–79511–6 ML2911 .D35 1979

     Study of the primitive multisectional New England anthem, approximately 1760–1800. Considers 118 indigenous American anthems with comparisons to more than seventy anthems by eighteenth-century English composers. Confirms the influence of an English singing-school movement on the New England movement. Works of twenty-one American composers, including William Billings and such lesser known figures as Jacob French, Daniel Read, Samuel Holyoke, William Cooper, Supply Belcher, and Oliver Holden. Two appendices: anthems by non-American composers published in New England before 1800; anthems by native composers published in New England before 1800. Many musical examples throughout the text;

lengthy bibliography lists not only books, articles, and manuscripts, but some 125 scores and collections of anthems; expansive index.

389. Fansler, Terry Lee. "The Anthem in America, 1900–1950." Ph.D. dissertation. Denton: University of North Texas, 1982. xvii, 230 p.

Study of anthem literature published and performed in the United States, 1900–1950. Focuses on the quartet anthem, anthems in the Episcopal tradition, prominent choral ensembles, dissemination of the anthem, anthems by prominent music educators, anthems in the Russian style, and the Negro spiritual. Biographical sketches of major composers and detailed analyses of selected anthems. Musical examples; bibliography of over 150 writings, editions of music, and recordings.

390. Wienandt, Elwyn A., and Robert H. Young. *The Anthem in England and America.* New York: The Free Press, 1970. xiii, 495 p. ML3265 .W53

First comprehensive study of the anthem since Myles B. Foster's long-outdated *Anthems and Anthem Composers* (1901). Major addition to the literature when originally published. Remains fine study of the anthem from Reformation to the 1960s. First four chapters provide a concise description of the origin and development of the anthem in England from the Reformation to 1825. Fifth and sixth chapters concern decorative church music in America, early eighteenth to mid-nineteenth century. In the rest of the book, alternate chapters trace developments in England and America through the 1960s. Significant Canadian contributions included. Numerous musical examples; valuable extensive bibliography; expansive index.

CAMP MEETING SONGS

391. Hulan, Richard Huffman. "Camp-Meeting Spiritual Folksongs: Legacy of the 'Great Revival in the West'." Ph.D. dissertation. Austin: University of Texas, 1978. xxxii, 246 p.

Study of the nineteenth-century religious folk song in the United States, focusing on the camp meeting songs published between 1800 and 1813. Annotated bibliography of 541 camp meeting songs published between 1800 and 1805. Maps and tables; various bibliographies throughout, including fifty-two printed sources containing texts of camp meeting songs and general bibliography of approximately 110 writings.

392. Johnson, Charles A. *The Frontier Camp Meeting: Religion's Harvest Time.* Dallas, Tex.: Southern Methodist University Press, 1955. ix, 325 p. BX8475 .J64; reprint, Dallas, Tex.: Southern Methodist University Press,

1985. New introduction by Ferenc M. Szasz. xxi, 325 p. ISBN 0–87074–201–9 BX8475 .J64 1985

The history of camp meetings in early-nineteenth-century trans-Allegheny West (portions of North Carolina, South Carolina, Virginia, Georgia, Alabama, the Ohio Valley, and Mississippi Valley). Chapter 10 is devoted to camp meeting hymns. See "camp meeting songs" in the expansive index for other locations in the text for references to the music. Illustrations; classified bibliography of approximately 250 writings, with thirty-four pertaining to revival hymnology and folk music.

393. Lorenz, Ellen Jane. *Glory, Hallelujah!: The Story of the Campmeeting Spiritual.* Nashville, Tenn.: Abingdon, 1980. 144 p. ISBN 0–687–14850–2 ML3186 .L86

Based in part on Lorenz's doctoral dissertation, "A Treasure of Campmeeting Spirituals" (Union Graduate School, 1978). Recounts the songs, performance practice, and culture of the nineteenth-century camp meetings. Demonstrates the persistence and influence of the camp meetings on present-day Protestant and Catholic music practices. Musical examples; glossary; bibliography of fifty-four writings; index.

394. Porter, Ellen Jane. "A Treasure of Campmeeting Spirituals." Ph.D. dissertation. Yellow Springs, Ohio: Union Graduate School, 1978. xii, 445 p.

Provides an overall history of camp meetings and revivals, with primary focus on camp meeting spirituals. Surveys collections of camp meeting spirituals, especially Joseph H. Hillman's *The Revivalist* (Hillman, 1868). Lengthy annotated bibliography of 607 camp meeting spirituals and general folk hymns, a majority drawn from *The Revivalist.* Musical incipits and index of titles included. Glossary of terms; chronology of nearly 250 significant music publications containing camp meeting songs; general bibliography of approximately 160 writings; bibliography of nearly two hundred collections containing camp meeting songs; bibliography of eighteen writings about *The Revivalist.*

CANATAS

See: Evans, Margaret R. *Sacred Cantatas: An Annotated Bibliography, 1960–1979* (item 82)

CHORAL MUSIC

395. Adler, Samuel, et al. *American Sacred Choral Music: An Overview and Handbook.* Foreword by Elizabeth Patterson. Introduction by Daniel Pinkham.

Brewster, Mass.: Paraclete Press, 2001. x, 77 p. ISBN 1–55725–276–9 ML2911 .A45 2001

Five essays by various authors. Essays: "*Harmonia Americana:* Our Legacy of Sacred Music" (C. Timberlake) surveys landmark music publications and historical studies; "Sacred Music in America: An Overview" (Adler); "The American Chorister" (Timberlake) identifies pedagogical resources for choral singing; "Performing the Music of Our Time" (J. E. Jordan, Jr.) offers suggestions for introducing twentieth-century sacred music to choirs; and "Music of the Synagogue" (Adler) discusses music of the synagogues in the United States. Musical examples and tables; bibliographies offered for some essays; an annotated bibliography/discography prepared by David Chalmers of twenty-two writings, twelve recordings, and a few Web sites.

396. Bruce, Neely. "Sacred Choral Music in the United States: An Overview" in *The Cambridge Companion to Singing,* ed. John Potter. Cambridge: Cambridge University Press, 2000 (rep. 2001). pp. 133–148. ISBN 0–521–62225–5, ISBN 0–521–62709–5 (pbk.) ML1460.C28 2000

Overview of sacred choral music in the United States intended for concert performance and for use in church services. Covers the latter half of the eighteenth century to the mid-twentieth century. Documented with endnotes; expansive index.

397. DeVenney, David P. *Source Readings in American Choral Music: Composers' Writings, Interviews & Reviews.* Missoula, Mont.: College Music Society, 1995. xiii, 258 p. ISBN 0–9650647–0–0, ISBN 0–9650647–1–9 (pbk.) ML1511 .D48 1995

Purpose: "to present important documents relating to the history and performance of choral literature written in the United States." Prefaced with a seven-page chronology of American choral music. Main body organized into three parts: (1) music before 1830; (2) music from 1830–1920; and (3) music since 1920. In all, thirty reprinted essays and articles by twenty-eight composers with annotations and comments. Attributes of documents chosen: "reflections on the nature and purposes of choral music by major contributors to the repertory, critical responses to landmark works, and instruction on performance practice." Many writings address sacred music. Classified bibliography of approximately 270 writings; expansive index.

398. DeVenney, David P. *Varied Carols: A Survey of American Choral Literature.* Westport, Conn.: Greenwood Press, 1999. xi, 315 p. ISBN 0–313–31051–3 ML1511 .D48 1999

Historical survey of sacred and secular American choral literature, based on examination of almost three thousand choral works by nearly three hundreed composers active in the United States between 1760 and the 1990s. Narrowly focused on the literature of American choral music, not on its creators, performers, or conductors. Provides brief descriptive analyses for some works. Bibliography of nearly three hundred writings; index.

399. Lundberg, John William. "Twentieth Century Male Choral Music Suitable for Protestant Worship." D.M.A. dissertation. Los Angeles: University of Southern California, 1974. iii, 178 p.

Broadly surveys male choral music with primary focus on twentieth-century choral compositions for adult male voices suitable for Protestant worship. Discusses representative compositions and presents critical analysis. Provides a list of eighty-five choral works for male voices. Bibliography of fifty-eight writings.

400. Rapp, Robert Maurice. "Stylistic Characteristics of the Short Sacred Choral Composition in the U.S.A., 1945–1960." Ph.D. dissertation. Madison: University of Wisconsin, 1970. ix, 298 p.

Analytical study of the "most frequently performed American sacred choral music for mixed voices" composed between 1945 and 1960. Examines text, arrangement of voices and instrument(s), musical characteristics, and idiomatic choral devices. Tables and musical examples; classified bibliography of about forty-five writings; list of publishers with addresses appended.

401. Reid, Robert Addison. "Russian Sacred Choral Music and Its Assimilation into and Impact on the American A Cappella Choir Movement." D.M.A dissertation. Austin: University of Texas, 1983. xvii, 264 p.

Traces the development of unaccompanied choral music of the Russian Orthodox church, 1650 to 1917, and its influence upon sacred music of the American a cappella movement of the first half of the twentieth century. Bibliography of about 120 items (majority in English).

402. Wienandt, Elwyn A. *Choral Music of the Church.* New York: The Free Press, 1965 (rep., New York: Da Capo Press, 1980). xi, 494 p. ISBN 0–306–76002–9 (DCP) ML3000 .W53

Three parts: (1) Catholic contribution, (2) Catholic traditions and Protestant innovation, and (3) the breakdown of denominational distinctions. Chronological coverage, Middle Ages to the present. Primary focus on European music, but includes discussion of the anthem, oratorio, Passion, and cantata in eighteenth- through twentieth-century America. Numerous musical examples, with facsimiles, tables, and photographs; general, annotated bibliography/discography of nearly 275 books, music editions, periodicals, articles, and recordings; expansive index.

See also: Bentley, Brian Richard. "The Philosophical Foundations and Practical Use of Choral Music in the Church of Jesus Christ of Latter-Day Saints in the Twentieth Century, Focusing on the Music of Robert Cundick, with Emphasis on His Sacred Service, *The Redeemer*" (item 357); Brown, Bruce Calvin. "The Choral Music of the American Lutheran Church: A Study and Performance of the Music of Daniel Moe, Paul Fetler, and Rolf Espeseth" (item 328); Carroll, Lucy E. "Three Centuries of Song: Pennsylvania's Choral Composers, 1681–1981" (item 194); Klassen, Roy Leon. "The Influences of Mennonite College Choral Curricula upon Music Practices in American Mennonite Churches" (item 334); Kroeger, Karl. "The Moravian Choral Tradition: Yesterday and Today" (item 352); Liemohn, Edwin. *The Organ and Choir in Protestant Worship* (item 221); McCalister, Lonnie. "Developing Aesthetic Standards for Choral Music in the Evangelical Church" (item 644); Stanislaw, Richard J. "Choral Performance Practice in the Four-Shape Literature of American Frontier Singing Schools" (item 466)

FOLK MUSIC

403. Jackson, George Pullen. *Another Sheaf of White Spirituals.* Foreword by Charles Seeger. Gainesville: University of Florida Press, 1952. xviii, 233 p. M2117 .A64 1952; reprint, New York: Folklorica, 1981. ISBN 0–939544–01–6 M1629 .A64 1981

Transcriptions of 363 American religious folksongs with historical commentary. Folksongs organized into nine groups: (1) songs which reveal the folk singing manner; (2) camp meeting spiritual songs; (3) songs which tell a religious story; (4) songs with tunes robbed from fiddlers, fifers, harpers, and frolickers; (5) folksongs of praise; (6) psalm tunes and others in a like mood; (7) sundry songs; (8) early lyric "lend-lease" to Britain; and (9) two folksongs and their "fuguing" transformations. Illustrations; bibliography of approximately 150 writings; index of secular songs and general index of titles.

404. Jackson, George Pullen. *Spiritual Folk-Songs of Early America: Two Hundred and Fifty Tunes and Texts, with an Introduction and Notes.* Preface by John Powell. 1st ed. New York: J. J. Augustin, 1937 (rep. 1938). x, 254 p. M1629.J147 S85; 2nd ed. Locust Valley, N.Y.: J. J. Augustin, 1953. M1629.J147 S85 1953; reprint of 1937 ed., New York: Dover, 1964. M1629.J147 S85 1964; reprint of 1937 ed., Gloucester, Mass.: Peter Smith, 1975. ISBN 0–8446–2297–4 M1629.J147 S85 1975

Transcriptions with historical commentary of fifty-one religious ballads, ninety-eight folk hymns, and 101 revival spiritual songs. Photos; bibliography of approximately 130 writings; index of song titles and index of first lines.

Jackson, George Pullen. *Down-East Spirituals and Others: Three Hundred Songs Supplementary to the Author's Spiritual Folk-Songs of Early America.* 1st ed. New York: J. J. Augustin, 1939 (rep. 1943). 296 p. M1629.J147 D6; 2nd ed. New York: J. J. Augustin, 1953. M1629.J147 D6 1953; reprint of 1943 ed., New York: Da Capo Press, 1975. ISBN 0–306–70666–0 M1629.J147 D6 1975

Transcriptions with historical commentary of sixty religious ballads, 152 folk hymns, and eighty-eight revival spiritual songs. Illustrations; bibliography of approximately 130 writings; index of song titles and index of first lines.

405. Jackson, George Pullen. *White and Negro Spirituals, Their Life Span and Kinship: Tracing 200 Years of Untrammeled Song Making and Singing among Our Country Folk: With 116 Songs as Sung by Both Races.* New York: J. J. Augustin, 1943. xiii, 349 p. ML3551 .J17; reprint, Locust Valley, N.Y.: J. J. Augustin, 1970. ML3551 .J17 1970; reprint, New York: Da Capo Press, 1975. ISBN 0–306–70667–9 ML3551 .J17 1975

Two parts: (1) history of religious folk song in the United States as sung by whites and (2) history of religious folk song as sung by African Americans. Discusses the "Great Awakening," music of the Baptists, Shakers, Mormons, and others, the spiritual, and camp meeting songs. Includes transcriptions with annotative remarks of 116 melodies sung by Whites paired with African-American variants, accompanied by an index to titles, first lines, and refrains. Musical examples, illustrations, photos, and tables; chronologically arranged bibliography of nearly 130 British and American books containing religious folk songs.

406. Jackson, George Pullen. *White Spirituals in the Southern Uplands: The Story of the Fasola Folk, Their Songs, Singings, and "Buckwheat Notes."* Chapel Hill: University of North Carolina Press, 1933. xv, 444 p. ML3551 .J2; reprint, Hatboro, Pa.: Folklore Associates, 1964. 444 p. ML3551 .J2 1964; reprint, New York: Dover, 1965. xvi, 444 p. ML3551 .J2 1965

American religious folksongs from their beginning in New England through their migration west and to southern regions of the United States. Musical examples, illustrations, and photos; bibliography of forty-six writings; expansive index.

407. Peterson, Robert Douglas. "The Folk Idiom in the Music of Contemporary Protestant Worship in America." Ed.D. dissertation. New York: Columbia University, 1972. 3, v, 236 p.

Examines the presence of folk elements in Protestant church music. Concentrates on music of the mid-twentieth century. Excludes jazz and rock idioms. Musical examples; bibliography of nearly two hundred writings and twenty scores.

See also: Epstein, Dena J. *Sinful Tunes and Spirituals: Black Folk Music to the Civil War* (item 230)

FUGING TUNES

408. Fawcett-Yeske, Maxine Ann. "The Fuging Tune in America, 1770–1820: An Analytical Study." Ph.D. dissertation. Boulder: University of Colorado, 1997. xvii, 563 p.

Traces the "evolution, development, and ultimate decline" of the fuging-tune in the United States during the time of it greatest popularity. Identifies British antecedents and examines tunes by approximately forty American composers with analyses. More than one hundred tunes shown in modern transcription. Bibliography of about 320 writings; index of tunes cross referenced to critical commentary; and, rare for a dissertation, a much appreciated expansive general index.

409. Temperley, Nicholas, and Charles G. Manns. *Fuging Tunes in the Eighteenth Century.* Detroit, Mich.: Information Coordinators, 1983. xi, 493 p. ISBN 0–89990–017–8 ML128.H8 T45 1983

History of British and American fuging tunes in the late seventeenth through eighteenth centuries. Bibliography of sources lists publications containing fuging tunes with English-language texts published 1700 to 1800, with 203 of the sources published in the United States. Inventories 1,239 fuging tunes found within the sources, supplying tenor incipit, meter, number of voices, structure, date, composer, source, key, reference letters and numbers, cross-references to the index of texts, tune name, and other descriptive comments. Bibliography of thirty-six secondary source writings; four indexes: (1) texts, (2) tune names, (3) persons, and (4) tunes in modern editions.

GOSPEL MUSIC

410. Allen, Ray. *Singing in the Spirit: African-American Sacred Quartets in New York City.* Philadelphia: University of Pennsylvania Press, 1991. xx, 268 p. ISBN 0–8122–3050–7, ISBN 0–8122–1331–9 (pbk.) ML3187 .A44 1991

Originated as the author's doctoral dissertation titled "Singing in the Spirit: An Ethnography of Gospel Performance in New York City's African-American Church Community" (University of Pennsylvania, 1987).

Historical study of African-American sacred quartets in New York City from the late nineteenth century to the present. Photographs; discography and videography of selected performances; bibliography of nearly two hundred writings; expansive index.

411. Boyer, Horace Clarence. *How Sweet the Sound: The Golden Age of Gospel.* Washington, D.C.: Elliott & Clark, 1995. 272 p. ISBN 1–880216–19–1 ML3187 .B7 1995; reprint, Urbana: University of Illinois Press, 2000. ISBN 0–252–06877–7 ML3187 .B7 2000

The 2000 reprint is published under the title *The Golden Age of Gospel.* Study of African-American sacred music, 1755 to the mid-1960s, in sacred and secular settings. Separate chapters on gospel music culture in specific U.S. cities and regions (Chicago, Philadelphia, Detroit, St. Louis, New York, and the state of Tennessee). Photographs; bibliography of approximately fifty writings; and index for song titles and an expansive index for subjects.

412. Broughton, Viv. *Black Gospel: An Illustrated History of the Gospel Sound.* Poole, Dorset, England: Blandford Press; New York: Distributed by Sterling, 1985. 160 p. ISBN 0–7137–1530–8, ISBN 0–7137–1540–5 (pbk.) ML3187 .B76 1985

Approximately 125 photos, illustrations, and facsimiles. Traces the history of African-American gospel music from slavery to the present. Naturally, covers a considerable amount of popular music. Index.

413. Burnim, Mellonee V. "The Black Gospel Music Tradition: Symbol of Ethnicity." Ph.D. dissertation. Bloomington: Indiana University, 1980. ii, 319 p.

Purpose: "to define the extent and the mechanisms through which gospel music serves as a symbol of ethnicity among Black Americans in the U.S." Musical examples and illustrations; bibliography of about 120 writings.

414. Burnim, Mellonee V. "The Performance of Black Gospel Music as Transformation" in *Music and the Experience of God,* ed. Mary Collins, David Power, and Mellonee Burnim. Edinburgh, Scotland: T. & T. Clark, 1989. pp. 52–61. ISBN 0–567–30082-X ML197 .M817 1989

Study "seeks to codify patterns of behavior which point to the existence of an underlying system of cultural values among Black Americans in the U.S." Documented with endnotes.

415. Cusic, Don. *The Sound of Light: A History of Gospel Music.* Bowling Green, Ohio: Bowling Green State University Popular Press, 1990. iv, 267 p. ISBN 0–87972–497–8, ISBN 0–87972–498–6 (pbk.) ML3187 .C88 1990

Traces the evolution of gospel music in the United States. Focuses on music of Protestant churches and on the commercial entertainment industry. Bibliography of more than three hundred writings; three indexes: subjects, proper names, and song titles.

416. DjeDje, Jacqueline Cogdell. "A Historical Overview of Black Gospel Music in Los Angeles." *Black Music Research Bulletin,* 10/1 (Spr. 1988): 1–5

Gospel music in Los Angeles from the 1920s to the 1980s. Discusses gospel music performed in churches, including some Catholic churches, and in secular settings. Bibliography of twenty-six writings.

417. Goff, James R. *Close Harmony: A History of Southern Gospel.* Chapel Hill: University of North Carolina Press, 2002. xiv, 394 p. ISBN 0–8078–2681–2, ISBN 0–8078–5346–1 (pbk.) ML3187 .G64 2002

Examines the American gospel music tradition. Part I covers gospel music and shape-note singing in nineteenth-century America. Parts II and III cover the birth of the gospel music industry, emergence of professional quartets, gospel music promoters, and the "All-Night Sings." Part IV covers the national expansion of gospel music and the emergence of southern gospel. Musical examples and photos; documented with endnotes; expansive index.

418. Heilbut, Anthony. *The Gospel Sound: Good News and Bad Times.* Updated, rev., 5th Limelight ed. New York: Limelight Editions, 1997. xxxv, 402 p. ISBN 0–87910–034–6 ML3187 .H44 1997

Originally published in 1971. History of the gospel movement in the United States. Photographs; classified, annotated discography of approximately 120 recordings; separate indexes for names and song titles.

419. Jackson, Irene V. "Afro-American Gospel Music and Its Social Setting: With Special Attention to Roberta Martin." Ph.D. dissertation. Middletown, Conn.: Wesleyan University, 1974. vi, 355 p.

Covers the African-American folk church and the function of gospel music apart from the worship service. Chapter 4 is a biography of Roberta Martin. Concludes with a catalog of approximately two thousand compositions by six hundred African-American gospel composers written between 1938 and 1965. Facsimiles, photographs, and musical examples.

420. Reagon, Bernice Johnson, ed. *We'll Understand It Better By and By: Pioneering African American Gospel Composers.* Washington, D.C.: Smithsonian Institution Press, 1992. xii, 384 p. ISBN 1–56098–166–0, ISBN 1–56098–167–9 (pbk.) ML390 .W274 1992

Essays/interviews about gospel composers, including Charles Albert Tindley, Lucie E. Campbell Williams, Thomas Andrew Dorsey, William Herbert Brewster, Sr., Roberta Martin, and Kenneth Morris. Essays include: "Pioneering African American Gospel Music Composers: A Smithsonian Institution Research Project" (Reagon); "The Impact of Gospel Music on the Secular Music Industry" (P. K. Maultsby); "Searching for Tindley" (Reagon); "Charles Albert Tindley: Progenitor of African American Gospel Music" (H. C. Boyer); "Lucie E. Campbell: Composer for the National Baptist Convention" (Boyer); "Lucie E. Campbell: Her Nurturing and Expansion of Gospel Music in the National Baptist Convention, U. S. A., Inc." (L. A. George); "Lucie E. Campbell Williams: A Cultural Biography" (C. Walker); "Take My Hand, Precious Lord, Lead Me On" (Boyer); "Conflict and Resolution in the Life of Thomas Andrew Dorsey" (M. W. Harris); "William Herbert Brewster: Rememberings" (Reagon); "William Herbert Brewster: The Eloquent Poet" (Boyer); "If I Fail, You Tell the World I Tried" (A. Heilbut); "William Herbert Brewster: Pioneer of the Sacred Pageant" (W. H. Wiggins, Jr.); "Roberta Martin: Spirit of an Era" (P. Williams-Jones); "Roberta Martin: Innovator of Modern Gospel Music" (Boyer); "Conversations: Roberta Martin Singers Roundtable" (Williams-Jones, Reagon); "Kenneth Morris: Composer and Dean of Black Gospel Music Publishers" (Boyer); and "Kenneth Morris: 'I'll be a Servant for the Lord'" (Reagon). Musical examples and photographs; bibliography of almost two hundred writings; classified discography of approximately 115 gospel and forty popular music recordings; classified, annotated bibliography prepared by Lisa Pertillar Brevard of nearly 120 writings; expansive index.

421. Reynolds, William Jensen. *Songs of Glory: Stories of 300 Great Hymns and Gospel Songs.* Grand Rapids, Mich.: Zondervan Books, 1990. 347 p. ISBN 0–310–51720–6 BV315 .R49 1990; reprint, Zondervan Books, 1995. 352 p. ISBN 0–8010–5527-X BV315 .R49 1995

Title self explanatory. Originated as weekly newspaper article series. Author explains that "the stories are intended to help make understandable the circumstances that produced these songs and the individuals who wrote, arranged, recorded, and preserved them." General index.

422. Tallmadge, William H. "The Responsorial and Antiphonal Practice in Gospel Song." *Ethnomusicology: Journal of the Society for Ethnomusicology,* 12/2 (May 1968): 219–238

Examines responsorial and antiphonal practice (beginning ca. 1870) in gospel song commonly sung in the United States. Theorizes as to the source of the practice. Numerous musical examples; bibliography of eighteen writings.

423. Williams-Jones, Pearl. "Afro-American Gospel Music: A Crystallization of the Black Aesthetic." *Ethnomusicology: Journal of the Society for Ethnomusicology,* 19/3 (Sept. 1975): 373–385

Discussion of Afro-American gospel music, its relationship to African heritage, and its importance for cultural identity. Bibliography of twenty-two writings and a discography of nine recordings.

See also: Collins, Mary, David Power, and Mellonee Burnim, eds. *Music and the Experience of God* (item 629); Dargan, William T. "Congregational Gospel Songs in a Black Holiness Church: A Musical and Textual Analysis" (item 314); Jackson, Irene V., ed. *More Than Dancing: Essays on Afro-American Music and Musicians* (item 231); Kaatrud, Paul Gaarder. "Revivalism and the Popular Spiritual Song in Mid-Nineteenth Century America: 1830–1870" (item 456)

HYMNS AND HYMNODY

General

424. Alexander, J. Neil, ed. *With Ever Joyful Hearts: Essays on Liturgy and Music: Honoring Marion J. Hatchett.* New York: Church Publishing, 1999. vi, 417 p. ISBN 0–89869–321–7 BX5940 .W58 1999

A collection of essays in honor of Marion J. Hatchett. Three essays relate to church music in the United States: "A Singing Clarity, a Steady Vision: Marion Hatchett's Work as a Teacher of Liturgical Music" (C. Doran) traces Hatchett's contribution to the music of the Episcopal Church; "Returning to Our Musical Roots: Early Shape-Note Tunes in Recent American Hymnals" (H. Eskew) surveys the inclusion of shape-note folk hymns in American hymnals, primarily hymnals used in the Methodist, Baptist and Southern Baptist, Lutheran, and Episcopal churches; and "Americans in the *English Hymnal* of 1906" (D. W. Music) recognizes the extensive use of text and music by authors and composers of the United States in the *English Hymnal* of 1906, a landmark British collection. Essays documented with endnotes; some tables.

425. Benson, Louis F. *The English Hymn: Its Development and Use in Worship.* New York: Hodder & Stoughton; Philadelphia: Presbyterian Board of Publication, 1915. xvii, 624 p. BV312 .B4; reprint, Richmond, Va.: John Knox Press, 1962. BV312 .B4 1962

Dated, but worthy contribution. Chronological treatment of the development and use of hymns and hymnals in Great Britain and the United States. Topics specifically related to U.S. church music include the influence of

Isaac Watts on American psalmody, the hymnody of American Methodism, evangelical hymnody in America, the Literary Movement in America, and Oxford influences on American hymnody. The expansive index can further direct one to many fruitful topics, such as "Baptists: United States," "children's hymns: American Sunday school," "Philadelphia: Moravians," and so on.

426. Caswell, Austin B. "Social and Moral Music: The Hymn" in *Music in American Society 1776–1976: From Puritan Hymn to Synthesizer.* Edited by George McCue. New Brunswick, N.J.: Transaction Books, 1977. pp. 47–71. ISBN 0–87855–209-X, ISBN 0–87855–634–7 (pbk.) ML200.1 .M9

The essay collection, *Music in American Society 1776–1976,* includes twelve essays that examine, according to George McCure, the "specifically native character of music originating in the United States." Caswell's essay focuses on American hymnody and its influence on the American spiritual and gospel music. Musical examples and facsimiles; documented with endnotes.

427. Cheek, Curtis Leo. "The Singing School and Shaped-Note Tradition: Residuals in Twentieth-Century American Hymnody." D.M.A. dissertation. Los Angeles: University of Southern California, 1968. iii, 302 p.

Identifies thirty-five tunes from the singing school and shaped-note tradition in use (as of 1968) in hymnals of major Protestant churches in the United States (American Baptist, Southern Baptist, Seventh-Day Adventist, Presbyterian, Protestant Episcopal, Evangelical and Reformed Church, Church of Jesus Christ of Latter-Day Saints, Lutheran Church-Missouri Synod, Assemblies of God, Methodist, Congregational Christian Church, Church of the Nazarene, and Lutheran Church in America). Provides historical information about the publication and use of the tunes during the eighteenth and early nineteenth centuries. Each tune presented in musical notation followed by information, when known, on original source of tune, composer, original text, author of text, original source of text, hymnals where located, and musical variants, along with additional comments as needed. Bibliography of approximately two hundred writings and hymnals.

428. Christ-Janer, Albert, Charles W. Hughes, and Carleton Sprague Smith. *American Hymns Old and New.* New York: Columbia University Press, 1980. xv, 838 p. ISBN 0–231–03458-X M2117 .A573

Hughes, Charles W. *American Hymns Old and New: Notes on the Hymns and Biographies of the Authors and Composers.* New York: Columbia University Press, 1980 (rep. 1991). x, 621 p. ISBN 0–231–04934-X ML3270 .H8

Volume 1: American hymns dating as far back as the seventeenth century and sixty contemporary hymns by commissioned poets and composers. Organized by centuries; each set of hymns preceded by a historical perspective. A majority of the newly commissioned works were produced in the late 1950s. Called a "historical singing book" by Hughes. Not intended for use in the modern church service. Concludes with five helpful indexes: first lines/titles, authors and composers, tunes, meters, and Bible verses. Volume 2: Organized into two sections: (1) discussion of hymns presented in Volume 1 and (2) brief biographical sketches of authors and composers. Concludes with a bibliography of over 150 items.

429. Cross, Virginia Ann. "The Development of Sunday School Hymnody in the United States of America, 1816–1869." D.M.A. dissertation. La.: New Orleans Baptist Theological Seminary, 1985. vii, 688 p.

Development of Sunday school hymnody in the United States, 1816 to 1869, with intention of establishing its relationship to gospel hymnody common in the 1870s. Examines words-only hymnbooks, tunebooks, and songbooks. Musical examples, facsimiles, and tables; chronological bibliography of 341 U.S. Sunday school hymnbooks, 1816–1869, with index of compilers; classified bibliography (church music history; hymnology and music history) of approximately 180 writings.

430. Crowder, William S. "A Study of Lined Hymnsinging in Selected Black Churches of North and South Carolina." Ed.D. dissertation. University of North Carolina at Greensboro, 1979. vii, 106 p.

A study of the hymn singing practice of "lining-out" within seventy-five Baptist and Methodist churches in Piedmont, North and South Carolina. Presents analyses of twenty-six hymns. Musical examples and maps; bibliography of around one hundred writings.

431. Downey, James Cecil. "Joshua Leavitt's *The Christian Lyre* and the Beginning of the Popular Tradition in American Religious Song." *Revista de Música Latino Americana/Latin American Music Review,* 7/2 (Fall/Win. 1986): 149–161

Origin of the religious popular musical tradition in the United States, which "gave to urban America the Sunday school hymn, the gospel songs and solos, flourishing music publishing houses, and . . . printed music, recordings, concert artists, and media productions [in popular music styles]." Facsimiles; documented with endnotes.

432. Eskew, Harry. "William Walker's *Southern Harmony:* Its Basic Editions." *Revista de Música Latino Americana,* 7/2 (Fall/Win. 1986): 137–148

Researches the history of William Walker's *Southern Harmony,* first published in 1835 and revised four times thereafter (1840, 1847, 1847, and 1854). Table of editions; documented with endnotes.

433. Eskew, Harry, and Hugh T. McElrath. *Sing with Understanding: An Introduction to Christian Hymnology.* 2nd ed. Nashville, Tenn.: Church Street Press, 1995. xii, 400 p. ISBN 0–8054–9825–7 BV310 .E75 1995

Organized into three parts: (1) "The Hymn in Perspective" studies hymns as poetry and music, as well as from a theological perspective; (2) "The Hymn in History and Culture" is a chronological survey of selected hymn texts and tunes widely used in American hymnals; and (3) "The Hymn in Practice" examines the function of hymns within the mission of the church. Musical examples and illustrations; classified bibliography of more than eight hundred writings and scores; several indices: (1) general, (2) biblical reference, (3) hymn title and first line, and (4) hymn tune.

434. Foote, Henry Wilder. *Three Centuries of American Hymnody.* Cambridge, Mass.: Harvard University Press, 1940. x, 418 p. ML3111.F6 T4; reprint, Hamden, Conn.: Archon Books, Shoe String Press, 1961 (rep. 1968). ML3111.F6 T4 1961, ML3111.F6 T4 1968

Originally published in 1940; several reprints. Study extends from 1640 to 1940, from the publication of *The Bay Psalm Book* to its three hundredth anniversary. Ten chapters: "The Heritage of English Psalmody," "The Reign of *The Bay Psalm Book,*" "The Revival of Singing in Eighteenth-Century New England," "The Early Hymns and Times of the German Settlers in Pennsylvania," "The Transition from Psalmody to Hymnody," "The Opening of a New Era, 1800–1830," "The Mid-Century Flood Tide, 1831–1865," "Hymns of the Last Third of the Nineteenth Century, 1866–1900," "Hymns of the Twentieth Century," and "Retrospect and Prospect." Two essays appended: "The Controversy over the Practice of 'Lining-Out' the Psalms" and "The Controversy at South Braintree over 'Regular' Singing." The 1968 reprint includes a new appendix, "Recent American Hymnody," reprinted from the *Papers on the Hymn Society,* 17 (1952), 419–441. Three indexes: (1) names and subjects, (2) index of psalm and hymn books, and (3) index of first lines of psalms, hymns, and spiritual songs.

435. Hawn, C. Michael. "The Consultation on Ecumenical Hymnody: An Evaluation of Its Influence in Selected English Language Hymnals Published in the United States and Canada since 1976." *The Hymn: A Journal of Congregational Song,* 47/2 (Apr. 1996): 26–37

Provides a brief historical overview of the Consultation on Ecumenical Hymnody (CEH) and its influence on English language hymnals published in the United States and Canada in last quarter of twentieth century. Addresses textual changes, new translations, archaic language, gender-inclusive language, tunes, gospel hymns, and twentieth-century hymns. Presents information in six tables, some of which are updated by Hawn's July, 1997 article, "The Tie that Binds: A List of Ecumenical Hymns in English Language Hymnals Published in Canada and the United States since 1976" (item 95): (1) alphabetical list of hymnals surveyed; (2) surveyed hymnals by date of publication and according to faith tradition; (3) number of CEH hymns found in surveyed hymnals; (4) alphabetical listing of hymns from the CEH (1976) as found in selected English language hymnals published in the United States and Canada since 1976; (5) frequency of appearance of hymns in selected English language hymnals published in the United States and Canada since 1976; and (6) CEH hymns written in the twentieth century organized by decade.

436. Leaver, Robin A., and James H. Litton, eds. *Duty and Delight: Routley Remembered: A Memorial Tribute to Erik Routley (1917–1982), Ministry, Church Music, Hymnody.* Carlton R. Young, executive ed. Carol Stream, Ill.: Hope Publishing; Norfolk, England: Canterbury Press, Norwich, 1985. xiv, 310 p. ISBN 0–916642–27–5 (Hope), ISBN 0–907547–48–6 (CP) ML55 .R66 1985

Fourteen essays by various authors evaluate Routley's contribution to ministry, church music, and hymnody. Essays that specifically relate to church music in the United States: "The Theological Character of Music in Worship" (Leaver) presents a study of music in worship drawing upon the writings of Routley and Paul Waitman Hoon; "Twentieth-Century American Church Music" (A. Wyton) surveys the contributions of composers, performers, and significant organizations to U.S. church music in the twentieth century; "Church-Music Education in American Protestant Seminaries" (P. W. Wohlgemuth) reviews commitment of seminaries to church music education; "The Hymn in Anglican Liturgy" (Litton) discusses the relationship between hymns and the liturgy; and "The Hymn Renaissance in the United States" (R. Schulz-Widmar) examines the "hymn explosion" and compares new hymn texts and music with older hymn texts and music. Musical examples, tables, and photographs; bibliography of writings by Routley; no index.

Leaver, Robin A. *The Theological Character of Music in Worship.* St. Louis, Mo.: Concordia, 1989. 22 p. ISBN 0–570–01–339–9 ML3001 .L3 1989

A reprint of Leaver's essay from *Duty and Delight.*

437. Mauney, Richard Steadman. "The Development of Missionary Hymnody in the United States of America in the Nineteenth Century." D.M.A. dissertation. Fort Worth, Tex.: Southwestern Baptist Theological Seminary, 1993. vii, 163 p.

 Development of missionary hymns, a body of congregational songs composed for use in the early-nineteenth-century foreign missions movement. Includes works by British and American composers. Focuses on music of four denominations instrumental in the missions movement: Congregationalists, Baptists, Presbyterians, and Methodists. Tables; chronological listing of missionary hymnals classified by denomination and number; list of selected missionary hymns with statistical information regarding frequency of occurrence in nineteenth-century collections; bibliography of approximately 150 writings.

438. McCutchan, Robert Guy. *Hymn Tune Names: Their Sources and Significance.* Nashville, Tenn.: Abingdon Press, 1957 (rep., St. Clair Shores, Mich.: Scholarly Press, 1974, 1976, 1977). 206 p. ISBN 0–403–07203–4 (SP, 1976) ML3186 .M22

 By a distinguished American hymnologist. Discusses hymn names and variant titles primarily of hymns of Great Britain and the United States. Includes alphabetical, amply-annotated list of tune names. Interesting reading. Three indexes: melodic index; index of a few names not included in the alphabetical list; and an index of first words or lines of hymns.

439. Music, David W. *Hymnology: A Collection of Source Readings.* Lanham, Md.: Scarecrow Press, 1996. xix, 235 p. ISBN 0–8108–3148–1 ML3000 .M87 1996

 Arranged chronologically within geographical divisions. Consists of seventy-two writings drawn from letters, diaries, and scholarly writings addressing various aspects of hymnology. All writings in English with translations provided for those originally in other languages. Begins with early church song and covers the Reformation, English hymnody, American hymnody, and Vatican Council II (1962–1965). Includes writings from the second century to the 1960s but offers few twentieth-century sources. Classified bibliography of approximately eighty-five writings; index.

440. Patrick, Millar. *The Story of the Church's Song.* Rev. for American use by James Rawlings Sydnor. Richmond, Va.: John Knox Press, 1962. 208 p. ML3186 .P35

 Survey of Christian hymnody, including discussion of metrical Psalms and Scottish paraphrases. First published in 1927; Sydnor's revision adds descriptive footnotes to the original text and includes an appendix,

"American Hymnody 1927–61." No musical examples; the reader is referred to selected hymnals. Original classified bibliography (1927) of nearly forty writings; revised classified bibliography of about eighty writings; expansive general index and index of titles.

441. Reeves, Marjorie, and Jenyth Worsley. *Favourite Hymns: 2000 Years of Magnificat.* London: Continuum, 2001. xvii, 206 p. ISBN 0–8264–4872–0 ML3086 .F38 2001

Examines the hymn texts of about eighty-five popular hymns. A majority of the composers, arrangers, and hymn writers discussed are not of the United States. Sections relating to music of the United States include the African-American spiritual, Shaker hymns, and individuals such as James Russell Lowell, Phillips Brooks, and John White Chadwick, to name a few. Book displays a decisive British slant, often relating the contribution of American hymn writers and composers to the music culture in Great Britain. Since many of the chosen hymns are widely sung in worship services in the United States, the study is useful for U.S. church musicians.

442. Reynolds, William Jensen, and Milburn Price. *A Survey of Christian Hymnody.* 4th ed. Rev. and enlarged by David W. Music and Milburn Price. Carol Stream, Ill.: Hope, 1999. xx, 327 p. ISBN 0–916642–67–4 ML3186 .R5 1999

The 1st (1963), 3rd (1987), and 4th (1999) editions are published under the title *A Survey of Christian Hymnody.* The 2nd edition (1978) published under the title *A Joyful Sound: Christian Hymnody.* A study of Christian hymnody from the early church through the twentieth century. Two chapters devoted to American hymnody from colonial to present times. Documented with endnotes. Includes a supplementary classified bibliography of about two hundred writings of which approximately ninety are related to American hymnody. As a supplement, provides the music of 139 illustrative hymns. Two indexes: index of illustrative hymns and general index.

443. Rogal, Samuel J. *A General Introduction to Hymnody and Congregational Song.* Philadelphia: American Theological Library Association; Metuchen, N.J.: Scarecrow Press, 1991. x, 324 p. ISBN 0–8108–2416–7 BV312 .R64 1991

Hymnody and congregational song of "the English-speaking Western world." Three of the ten chapters, a total of 101 pages, are specifically devoted to hymnody and the African-American spiritual in the United States. Musical examples and tables; general bibliography of forty-five writings; index to persons, hymns, hymn collections, and tunes.

444. Routley, Erik. *Christian Hymns Observed: When in Our Music God is Glorified.* Foreword by Ray Robinson. Princeton, N.J.: Prestige Publications, 1982. vii, 121 p. ISBN 0–911009–00–0 (pbk.) ML3270 .R67 1982; London: Mowbray, 1983. ix, 121 p. ISBN 0–264–66893–6 ML3270 .R67 1983

Hymnody from the early Christian church to the 1970s. Several chapters touch on music in the United States, most notably, "Varied Scenery: Mainly Welsh and American, 19th Century," "The Crisis of Denial: The Popular Rebellion from 1955," "The Crisis of Custom: Roman Catholic Hymnody Since 1964," and "The Main Stream Since 1955." Sparsely documented with footnotes; index of hymns in American and British hymnals.

445. Routley, Erik. "Music in Christian Hymnody." Ph.D. dissertation. Oxford, England: Mansfield College, Oxford University, 1951. x, 769, 84 p.

Primary focus on English Protestant hymnody since the Reformation. Chapter 33 briefly covers American hymnody, including its background, early American tunes, Lowell Mason and his contemporaries, and modern American hymnody. Musical examples; classified bibliography of 136 writings and hymnals.

Routley, Erik. *The Music of Christian Hymnody: A Study of the Development of the Hymn Tune since the Reformation, with Special Reference to English Protestantism.* London: Independent Press, 1957. vii, 308 p. ML3186 .R73

Abridgement of his doctoral dissertation, "Music in Christian Hymnody." Also includes a section on American hymnody, but much less information than in the dissertation. Musical examples; index of musical selections included in the *English Hymnal*; general index.

446. Routley, Erik. *The Music of Christian Hymns.* Chicago.: G. I. A. Publications, 1981. 184, [200], 19 p. ISBN 0–941050–00–9 ML3000 .R68 1981

Study of hymnody from the early church to the twentieth century. Chapters 19 and 20 treat American hymnody, discussing the New England style, American folk hymnody, the African-American spiritual, gospel song, and ethnic hymnody brought to the United States by immigrants. Chapter 26 addresses twentieth-century hymnody in the United States, Canada, and Australia. Impressive number of musical examples; classified bibliography of approximately seventy-five writings and 225 hymnals and companions to hymnals; index of hymn tunes and general index.

447. Sallee, James. *A History of Evangelistic Hymnody.* Foreword by Jack Hyles. Grand Rapids, Mich.: Baker Book House, 1978. 103 p. ISBN 0–8010–8111–4 ML3186 .S18

Evangelistic hymnody in the United States from the eighteenth century to 1975. Discusses congregational song, revival hymns, the singing school movement, religious folk music, social hymnbooks, gospel song, and other topics. Musical examples; bibliography of forty-eight writings and five hymnals.

448. Sydnor, James Rawlings. *Hymns: A Congregational Study.* Carol Stream, Ill.: Agape, 1983. vii, 81 p. (student's ed.); iii, 19 p. (teacher's ed.) ISBN 0–916642–19–4 ML3001.S92 H95 1983, ML3001.S92 H951 1983

Student's edition: Addresses hymn texts, correlation of text and music, the organization of hymnals, history of congregational song, and developing congregational singing. Illustrations and musical examples; classified bibliography of nearly seventy writings, seventeen hymnals, and twelve periodical titles; index.

Teacher's edition: Provides suggestions for teaching topics addressed in student's edition.

449. Wallace, Robin Knowles. *Moving toward Emancipatory Language: A Study of Recent Hymns.* Foreword by Heather Murray Elkins. Lanham, Md.: Scarecrow Press, 1999. xii, 269 p. ISBN 0–8108–3640–8 BV312 .W37 1999

Originated as the author's doctoral dissertation (Northwestern University, 1996). Study "explores a strand of English-language hymn texts [represented by Isaac Watts and Charles Wesley] intended for congregational singing that, generally speaking, have poetic qualities and are strophic, based on rhyme and meter." Chosen for closer evaluation: six hymn texts that exhibit elements of emancipatory language (referring in part to the absence of gendered terminology). Bibliography of approximately 270 writings; expansive general index, index of hymn texts and tunes, and index of biblical citations.

450. Wicker, Vernon, ed. *The Hymnology Annual: An International Forum on the Hymn and Worship.* Berrien Springs, Mich.: Vande Vere Publications, 1991–1993. 3 vols.: xvi, 394 p.; xii, 226 p.; ix, 203 p. ISSN 1054–7495 (set), ISBN 0–9628916–7–3 (vol. 2), ISBN 1–883218–05–5 (vol. 3) ML3270 .H9562

International in scope. Each volume contains nearly twenty essays by various authors. Documented with endnotes; some essays include bibliographies. Musical examples, facsimiles, illustrations, and photos included in some essays.

Volume 1: selective list of essays relating to church music in the United States: "'If They Don't Sing It, They Don't Believe It.' Singing in the

Worship Service. Expression of Faith or Unreasonable Liturgical Demand?" (P. Harnoncourt) raises the question, "Why should we sing as a church?"; "Theological Considerations for Poetic Texts Used by the Assembly" (T. H. Troeger); "Poet in the Congregation" (B. Wren) discusses the hymn in relationship to the contemporary church; and "Church Music in the U.S.A.: Steps toward Renewal of the Worship Services" (Wicker).

Volume 2: selective list of essays: "Sacred Sound and Meaning: Theological Reflections on Music and Word in Christian Worship" (R. Viladesau) examines the place of music in Christian worship from a Catholic perspective; "The Use and Performance of Hymnody, Spirituals, and Gospels in the Black Church" (P. K. Maultsby); "What is a Good Hymn?" (G. Aeschbacher) provides criteria for evaluating church music; "Metrical Psalmody: A Tale of Two Traditions" (E. R. Brink) recounts the author's experience with metrical Psalm and hymn singing in the Christian Reformed tradition in North America; and "Congregational Psalmody According to the Lutheran Tradition: Spiritually, Musically, Liturgically" (F. Schulz) primarily discusses the evolution of psalmody in the Lutheran tradition in Europe, but includes discussion of the relationship of the European Lutheran tradition to music included in *Lutheran Book of Worship* (Augsburg Publishing House; Board of Publication, Lutheran Church in America, 1978).

Volume 3: selective list of essays: "Beyond 'Alternative' and 'Traditional' Worship" (P. Westermeyer) identifies seven characteristics of worship rather than debates between the two approaches to worship; "Walking by Faith. An Exploration of the Devotional Context in Which the Hymn and Worship-Song are Used" (A. Luff), "Music and the Liturgy" (R. Warren), and "The Language of Hymns: Some Contemporary Problems" (J. R. Watson), while not specifically about U.S. church music, the message of these essays is universal; "Swedish Tradition in Swedish American Hymnals and Songbooks" (G. Grindal); "Gregorian Chant in Our Time" (Harnoncourt), although the author relies on music examples with German text, principles addressed may be applied to other vernacular languages; and "Biblical Text and Metaphor in Charles Wesley's Hymns" (Watson) examines Wesley's use of scripture and metaphor in his hymns.

451. Witvliet, John D. "The Blessing and Bane of the North American Megachurch: Implications for the Twenty-First Century Congregational Song." *The Hymn: A Journal of Congregational Song,* 50/1 (Jan. 1999): 6–14

Describes six characteristics of congregational song common to huge evangelical churches. Documented with footnotes; classified bibliography of sixty-five resources on contemporary worship and large evangelical churches.

452. Young, Carlton R. *My Great Redeemer's Praise: An Introduction to Christian Hymns.* Akron, Ohio: OSL Publications, 1995. ii, 171 p. ISBN 1–878009–22–2 BV310 .Y68 1995

Study of Christian hymns, hymn singing, and hymnals in the United States. Bibliography of about seventy writings; tunes index, hymns index, and expansive general index.

See also: Baldridge, Terry L. "Evolving Tastes in Hymntunes of the Methodist Episcopal Church in the Nineteenth Century" (item 342); Boccardi, Donald. *The History of American Catholic Hymnals: Since Vatican II* (item 263); Buehler, Kathleen D. *Heavenly Song: Stories of Church of God Song Writers and Their Songs* (item 34); Erickson, J. Irving. *Twice-Born Hymns* (item 385); Eskew, Harry, David W. Music, and Paul A. Richardson. *Singing Baptists: Studies in Baptist Hymnody in America* (item 250); Faus, Nancy Rosenberger, ed. *The Importance of Music in Worship* (item 333); Glover, Raymond F., ed. *The Hymnal 1982 Companion* (item 305); Gregory, David Louis. "Southern Baptist Hymnals (1956, 1975, 1991) as Sourcebooks for Worship in Southern Baptist Churches" (item 251); Guthrie, Joseph R. "Pentecostal Hymnody: Historical, Theological, and Musical Influences" (item 364); Hall, Roger L. "Shaker Hymnody: An American Communal Tradition" (item 379); Helseth, David C. "The Changing Paradigm of Congregational Music: A Disciples of Christ Response" (item 295); Hicks, Michael. *Mormonism and Music: A History* (item 359); Higginson, J. Vincent. *History of American Catholic Hymnals: Survey and Background* (item 282); Hinks, Donald R. *Brethren Hymn Books and Hymnals, 1720–1884* (item 260); Kindley, Carolyn E. "*Miriam's Timbrel:* A Reflection of the Music of Wesleyan Methodism in America, 1843–1899" (item 337); Moody, Michael Finlinson. "Contemporary Hymnody in the Church of Jesus Christ of Latter-Day Saints" (item 360); Murrell, Irvin Henry. "An Examination of Southern Ante-Bellum Baptist Hymnals and Tunebooks as Indicators of the Congregational Hymn and Tune Repertories of the Period with an Analysis of Representative Tunes" (item 255); Nordin, John P. "'They've Changed the Hymns!' or, An Investigation of Changes in Hymnody in the Hymnbooks of the Evangelical Lutheran Church in America" (item 310); Patterson, Beverly Bush. *The Sound of the Dove: Singing in Appalachian Primitive Baptist Churches* (item 369); Revell, Roger A. *Hymns in Worship: A Guide to Hymns of the Saints* (item 572); Reynolds, William Jensen. *Baptist Hymnody in America: From Roger Williams to Samuel Francis Smith* (item 256); Reynolds, William Jensen. *Songs of Glory: Stories of 300 Great Hymns and Gospel*

Songs (item 421); Schaeffer, Vicki. "An Historical Survey of Shaker Hymnody Expressing the Christian Virtues of Innocence and Simplicity" (item 382); Schalk, Carl. *God's Song in a New Land: Lutheran Hymnals in America* (item 330); Spencer, Jon Michael. *Black Hymnody: A Hymnological History of the African-American Church* (item 239); Spencer, Jon Michael. *Sing a New Song: Liberating Black Hymnody* (item 241); Stevenson, Arthur Linwood. *The Story of Southern Hymnology* (item 223); Stulken, Marilyn Kay. *Hymnal Companion to the Lutheran Book of Worship* (item 331); Tanner, Donald Ray. "An Analysis of Assemblies of God Hymnology" (item 248); Volland, Linda L. "Three Centuries of Methodist Hymnody: An Historical Overview of the Development of the American Methodist Hymnal with Special Attention to Hymnody in the 1780, 1878, and 1989 Hymnals" (item 340)

Revivalism

453. Downey, James Cecil. "The Music of American Revivalism." Ph.D. dissertation. New Orleans: Tulane University, 1968. viii, 219 p.

Purpose: "to isolate and to categorize the music influenced by American revivalistic activity in the period 1740 to 1800; to define the relationships between this music and the forces of social and religious change; and to formulate and present a new perspective of its historical value." Several useful appendixes, including a chronological listing of the publications of Isaac Watts, George Whitefield, John Newton, and John Rippon, and a chronological listing of American hymnals containing texts from the Separatist tradition. Musical examples; bibliography of nearly one hundred writings.

454. Frankiel, Tamar. *Gospel Hymns and Social Religion: The Rhetoric of Nineteenth-Century Revivalism.* Philadelphia: Temple University Press, 1978. xi, 222 p. ISBN 0–87722–142–1 BV460 .F7 1978

Published under the author's name of Sandra Sue Sizer, now known as Tamar Frankiel. Based on the author's doctoral dissertation, "Revival Waves and Home Fires: The Rhetoric of Late Nineteenth-Century Gospel Hymns" (University of Chicago, 1976). Early chapters survey the contributions of Dwight Lyman Moody (1837–1899) and Ira David Sankey (1840–1908) to American evangelism and hymnody. Primary focus on historical reconstruction of revivals and gospel-hymn rhetoric. A few tables; documented with endnotes; expansive index of personal names and subjects and an index of hymn titles sited in the study.

455. Hammond, Paul Garnett. "Music in Urban Revivalism in the Northern United States, 1800–1835." D.M.A. dissertation. Louisville, Ky.: Southern Baptist Theological Seminary, 1974. viii, 186 p.

Study of the role of music within "urban revivalism," a religious revival of the first quarter of the nineteenth century. Three influential hymnbooks are examined: Asahel Nettleton's *Village Hymns for Social Worship,* Joshua Leavitt's *The Christian Lyre,* and Lowell Mason's and Thomas Hastings' *Spiritual Songs for Social Worship.* Musical examples; classified bibliography of nearly ninety writings and thirty-five hymn and tunebooks; bibliography of approximately 250 hymnals published in the Northern United States between 1801 and 1830, which serve as sources for revival hymnody.

456. Kaatrud, Paul Gaarder. "Revivalism and the Popular Spiritual Song in Mid-Nineteenth Century America: 1830–1870." Ph.D. dissertation. Minneapolis: University of Minnesota, 1977. xi, 370 p.

Examines Protestant church music in mid-nineteenth-century United States, specifically the traditional revival song, American (or Lowell Mason-type) hymn tune, and gospel song. Musical examples; bibliography of approximately ninety writings.

Sacred Harp

457. Bealle, John. *Public Worship, Private Faith: Sacred Harp and American Folksong.* Athens: University of Georgia Press, 1997. xv, 308 p. ISBN 0–8203–1921-X, ISBN 0–8203–1988–0 (pbk.) ML3188 .B43 1997

Surveys *Sacred Harp* shape note singing, primarily in the Southern states. Organized into four chapters: (1) "Timothy Mason in Cincinnati: Music Reform on the Urban Frontier;" (2) "Sacred Harp as Cultural Object;" (3) "Writing Traditions of *The Sacred Harp*;" and (4) "'Our Spiritual Maintenance Has Been Performed': Sacred Harp Revival." Twelve appendixes comprising more than one hundred pages; bibliography of approximately 250 writings; expansive index.

458. Cobb, Buell E. *The Sacred Harp: A Tradition and Its Music.* Updated ed. Athens: University of Georgia Press, 1989. xvii, 245 p. ISBN 0–8203–0426–3, ISBN 0–8203–1022–0 (pbk.) ML311 .C6 1989

Originally published in 1978. Draws on scholarship of George Pullen Jackson (1874–1953). Reviews *Sacred Harp* singing traditions of various groups, with specific focus on the "Denson-book" singers. Appendix provides a listing with date and location of annual *Sacred Harp* "sings," though some information may not be reliable. For another source on *Sacred Harp* singing sessions, consult *Fasola* (item 721). Musical examples, facsimiles, and photos; bibliography of forty writings; expansive index.

459. Horn, Dorothy D. *Sing to Me of Heaven: A Study of Folk and Early American Materials in Three Old Harp Books.* Gainesville: University of Florida Press, 1970. xi, 212 p. ISBN 0–8130–0293–1 ML3111 .H67

"Old Harp" in the title refers to shape note gospel singing, where singers are often referred to as "Old Harp Singers." Historical and analytical study of music contained in *The Southern Harmony* (1835), *The Original Sacred Harp* (1836), and *The New Harp of Columbia* (1867). Part of the study is drawn from the author's earlier writings, specifically "Quartal Harmony in the Pentatonic Folk Hymns of the Sacred Harps," *Journal of American Folklore,* 71/282 (Oct.–Dec. 1958), 565–581 (also reprinted as a monograph: American Folklore Society, 1958) and the author's dissertation, "A Study of the Folk-Hymns of Southeastern America" (University of Rochester, 1953). Numerous musical examples; bibliography of nearly sixty writings and one hundred hymnals and singing school manuals; index of tunes; expansive general index.

460. Loftis, Deborah C. "Big Singing Day in Benton, Kentucky: A Study of the History, Ethnic Identity and Musical Style of *Southern Harmony* Singers." Ph.D. dissertation. Louisville: University of Kentucky, 1987. vii, 264 p.

History of "Big Singing Day," the fourth Sunday in May, held in Benton, Kentucky, where people gather to sing hymns from *The Southern Harmony* shape note hymnal. Presents analysis of core repertoire of *The Southern Harmony* and discusses performance practice, drawing a comparison between the musical style of *The Southern Harmony* singing to *The Sacred Harp* singing. Musical examples, facsimiles, photos, and tables; bibliography of about one hundred writings and forty field recordings.

See also: Ellington, Charles Linwood. "The *Sacred Harp* Tradition of the South: Its Origin and Evolution" (item 209)

Tunebooks

461. Bean, Shirley Ann. "*The Missouri Harmony,* 1820–1858: The Refinement of a Southern Tunebook." D.M.A. dissertation. University of Missouri-Kansas City, 1973. xii, 308 p.

Examines the significant nineteenth-century southern U.S. tunebook along with "northern" influences introduced through later revision by Charles Warren. Also includes a nineteen-page chapter on singing schools in the

United States. Several helpful indexes. Bibliography of approximately 175 writings and 30 tunebooks.

462. Crouse, David L. "The Work of Allen D. Carden and Associates in the Shape-Note Tune-Books *The Missouri Harmony, Western Harmony,* and *United States Harmony.*" D.M.A. dissertation. Louisville, Ky.: Southern Baptist Theological Seminary, 1972. iv, 192 p.

    Examines the content of three significant U.S. shape note tunebooks, as indicated in the title. Provides biographical information for their compiler, Allen D. Carden and associates, Samuel J. Rogers, F. Moore, J. Green, and Charles Warren. Musical examples and facsimiles; bibliography of more than one hundred writings and forty tune and hymnbooks. As an appendix, an alphabetical list of tunes with composers and authors.

463. Eskew, Harry. "Shape-Note Hymnody in the Shenandoah Valley, 1816–1860." Ph.D. dissertation. New Orleans: Tulane University, 1966. xi, 170 p.

    The Shenandoah Valley, Virginia, was an important "center of activity of the American singing-school movement that spread from New England to the South and West." Study extends from the beginning of the use of shape note hymnody to 1860. Includes the tunebooks of Ananias Davisson, Joseph Funk, James P. Carrell, David L. Clayton, among others. Musical examples, illustrations, and tables; chronological list of Shenandoah Valley tunebooks; bibliography of approximately ninety writings and nearly seventy tunebooks and hymnals.

464. Kroeger, Karl. "*The Worcester Collection of Sacred Harmony* and Sacred Music in America, 1786–1803." Ph.D. dissertation. Providence, R.I.: Brown University, 1976. xxxiii, 675 p.

    Biography of *The Worcester Collection of Sacred Harmony,* one of the most popular and influential tunebooks in the United States during the late eighteenth and early nineteenth centuries. Numerous musical examples and tables; bibliography of nearly three hundred tunebooks, thirty-nine hymnals and collections of verse, eleven hymnal companions, thirteen newspaper titles, and more than one hundred writings; several appendixes; expansive index.

465. Norton, Kay. *Baptist Offspring, Southern Midwife: Jesse Mercer's* Cluster of Spiritual Songs *(1810): A Study in American Hymnody.* Warren, Mich.: Harmonie Park Press, 2002. xxiii, 202 p. ISBN 0–89990–109–3 ML3160 .N67 2002

    Examines *Cluster of Spiritual Songs,* a text-only tunebook, in relationship to eighteenth-century sacred music in Georgia and the emergence of shaped-note notation common in the South. Musical examples, facsimiles, and tables; bibliography of about 160 writings and ninety hymnals and musical sources; three indexes: first line of the 1810 *Cluster* hymns, hymn tunes cited in the text or tables, and an expansive general index.

466. Stanislaw, Richard J. "Choral Performance Practice in the Four-Shape Literature of American Frontier Singing Schools." D.M.A. dissertation. University of Illinois at Urbana-Champaign, 1976. x, 468 p.

Performance practices of three- and four-voice music printed in four-shape notation in the United States during the nineteenth century. The tunebooks often contain theory and performance instructions and were primarily used at singing schools and for Protestant church functions. Covers the geographical areas of western Pennsylvania, Ohio, Kentucky, some of the Midwest, and most of the South. Musical examples; bibliography of approximately seventy writings and an annotated list of around 250 hymnbooks in four-shape notation.

See also: Davenport, Linda Gilbert. *Divine Song on the Northeast Frontier: Maine's Sacred Tunebooks, 1800–1830* (item 203); Graham, Fred Kimball. "*With One Heart and One Voice": A Core Repertory of Hymn Tunes Published for Use in the Methodist Episcopal Church in the United States, 1808–1878* (item 343)

INSTRUMENTAL MUSIC

467. Anderson, Mark J. *A Sourcebook of Nineteenth-Century American Sacred Music for Brass Instruments.* Westport, Conn.: Greenwood Press, 1997. vi, 130 p. ISBN 0–313–30380–0 ML933 .A53 1997

Aims to acquaint musicians with a body of sacred music written and arranged for brass instruments in the United States. Brief summary of the music of the Moravian, Evangelist, and Harmony Society (Harmonists) churches, as well as a discussion of other significant musicians. Sections on cornet music, small ensembles, large ensembles, and solos. Numerous musical examples, photos, illustrations, and facsimiles; bibliography of nearly fifty writings; index.

468. Kroeger, Karl. "The Church-Gallery Orchestra in New England." *The American Music Research Center Journal,* 4 (1994): 23–30

Describes the use of instruments in combination with vocal parts in eighteenth- and nineteenth-century New England worship services. Descriptions of musical practices supported by period diary writings. Although instrumental parts were rarely notated, gallery orchestras were formed, consisting of any number and combination of instruments and principally used to support the vocal parts. Documented with endnotes.

469. Music, David W. *Instruments in Church: A Collection of Source Documents.* Foreword by Robin A. Leaver. Lanham, Md.: Scarecrow Press, 1998. xix, 211 p. ISBN 0–8108–3595–9 ML3001 .I57 1998

Writings on the use of musical instruments in Jewish and Christian services. Organized chronologically. Within the eighteenth and nineteenth centuries, three sections relate to U.S. church music: "The Bass Viol and Gallery Orchestra in England and America," "The Introduction of the Organ into American Churches," and "The Churches of Christ Reject the Use of Instruments." Within the twentieth century, four sections relate to U.S. church music: "The Piano in American Musical Evangelism," "Roman Catholic Pronouncements on Instrumental Music," "The Use of Musical Instruments from the Folk, Rock, and Pop Cultures," and "Electronic Instruments in the Church." Musical examples; classified bibliography of approximately sixty writings; index.

470. Ode, James A. *Brass Instruments in Church Services.* Minneapolis, Minn.: Augsburg, 1970. 76 p. ML3111 .O34

Three parts: (1) overview of use of brass instruments in worship since 1750, specifically the musical practices in early Lutheran churches, New England, and nineteenth- and twentieth-century United States; (2) rehearsal for church performances; and (3) scoring for bass instruments. Musical examples and illustrations; three hymn arrangements for brass quintet appended; bibliography of nineteen collections of brass ensemble music suitable for church use; bibliography of nearly fifty writings.

471. Stormont, Beth Loreen. "Music for Organ and Instruments Written by American Composers since 1950 and Their Application in Protestant Worship." D.M.A. dissertation. Los Angeles: University of Southern California, 1978. iv, 197 p.

Chapter 2 offers a history of the use of instruments with organ in worship. Followed by the analysis of ten compositions for instruments with organ by American composers: Lloyd Pfautsch's *Festival Prelude on "I'll Praise My Maker,"* Richard Peek's *Pastoral and Noel*, Paul E. Koch's *Canon and Variations on "In Dulci Jubilo,"* Harald Rohlig's *Concertino Sacra on "Good Christian Men Rejoice,"* Leland B. Sateren's *Two Pieces for Organ and Brass*, Paul Creston's *Meditation*, Leo Sowerby's *Ballade*, Rayner Brown's *Poem*, Daniel Pinkham's *The Shepherd's Symphony*, and Robert Karlén's *Exclamation!* As an appendix, includes a bibliography of nearly 140 published compositions appropriate for use in Protestant worship, grouped by instrumental combinations; also lists the published compositions alphabetized by composer with additional information, including level of difficulty, descriptive category (based on preexistent melody; based on original theme(s); large or multimovement work; short work or freely composed prelude; festive work; meditative or more quiet work; varied work), recommended use (prelude; offertory; postlude; processional; recessional; introit), and publisher. Musical examples; bibliography of approximately 120 writings.

472. Trobian, Helen Reed. *The Instrumental Ensemble in the Church.* New York: Abingdon Press, 1963. 96 p. ML3001 .T86

Presents historical information on the use of instruments in worship, cites biblical references that describe instrumental music, and offers advice toward the effective use of instrumental music in worship. Five chapters: "Ensembles in Worship," "Ensembles in Christian Education," "Small Ensembles: String and Wind," "Small Ensembles: Brass," and "Large Ensembles." Bibliographies of recommended music throughout; expansive index.

MASSES

See: Foley, Edward, and Mary E. McGann. *Music and the Eucharistic Prayer* (item 273)

MOTETS

See: Balshaw, Paul. "A Selected Repertoire of Anthems and Motets for American Protestant Liturgical Churches" (item 80)

OPERAS

473. Smedley, Bruce Robert. "Contemporary Sacred Chamber Opera: A Medieval Form in the Twentieth Century." Ph.D. dissertation. Nashville, Tenn.: George Peabody College for Teachers, 1977. vii, 225 p.

Study of musical drama deemed appropriate for church performance, including medieval liturgical drama and the fourteenth-century mystery play, with primary focus on late-twentieth-century sacred chamber operas. Most works analyzed are by British and American composers. Musical examples and illustrations; bibliography of approximately sixty writings and forty scores and a discography listing a few recordings.

ORATORIOS

474. Smither, Howard E. *The Oratorio in the Nineteenth and Twentieth Centuries.* Chapel Hill: University of North Carolina Press, 2000. xxiv, 829 p. ISBN 0–8078–2511–5 ML3201 .S6 v.4

The oratorio in nineteenth- and twentieth-century Germany, England, France, and the United States. For nineteenth-century American oratorio: Chapter 7, "Oratorio in America: Cultural Context, Aesthetic Theory and Criticism" and Chapter 8, "Oratorio in America: Libretto, Music, Selected Oratorios." Does not offer separate chapters about the oratorio in twentieth-century America. The expansive subject index will prove helpful in this regard (see subject: "United States: church music in"). Musical examples, facsimiles, illustrations, and photos; bibliography of more than three hundred twentieth-century oratorios, some by U.S. composers; general bibliography of nearly six hundred writings and seven hundred oratorios.

ORGAN MUSIC

475. Arnold, Corliss Richard. *Organ Literature: A Comprehensive Survey.* 3rd ed. Metuchen, N.J.: Scarecrow Press, 1995. 2 vols.: iv, 391 p.; iv, 915 p. ISBN 0–8108–2970–3 ML600 .A76 1995

In two parts: (1) historical survey and (2) biographical catalog. Includes a twenty-four-page chapter on organ music in the United States from 1700 to 1994 with many works appropriate for church. Lists U.S. organ composers born between 1737 and 1960. Tables; bibliography of nearly eighty writings. Biographical catalog provides information about organ composers and their published works. Historical survey can be used in conjunction with the biographical catalog to locate entries on U.S. organ composers.

476. Kratzenstein, Marilou. *Survey of Organ Literature and Editions.* Ames: Iowa State University Press, 1980. x, 246 p. ISBN 0–8138–1050–7 ML600 .K73

Originally appeared as a series of articles published in *The Diapason* between 1971 and 1977. International in scope. One twenty-six-page chapter traces the sacred and secular history of organ music in the United States from Colonial times to the 1960s. Provides a bibliography of twelve collections of organ works and individual works by more than one hundred composers. Numerous musical examples; general bibliography of approximately four hundred writings; index of names and expansive index of subjects.

477. Ochse, Orpha Caroline. *The History of the Organ in the United States.* Bloomington: Indiana University Press, 1975. xv, 494 p. ISBN 0–253–32830–6, ISBN 0–253–20495-X (pbk.) ML561 .O3

The organ in the United States from 1524 to the early 1970s. Discusses organ building, sacred and secular use of organs, and the orchestral, American classic, and neobaroque organ. Photographs and tables; bibliography of 524 writings; expansive general index.

See also: Duncan, Timothy Paul. "The Role of the Organ in Moravian Sacred Music between 1740–1840" (item 345); Hieronymus, Bess. "Organ Music in the Worship Service of American Synagogues in the Twentieth Century" (item 319); Liemohn, Edwin. *The Organ and Choir in Protestant Worship* (item 221); Stormont, Beth Loreen. "Music for Organ and Instruments Written by American Composers since 1950 and Their Application in Protestant Worship" (item 471)

PSALMS AND PSALMODY

478. Clark, Linda J. *The Sound of Psalmody.* New York: Auburn Theological Seminary, 1975. 28 p. ML3011.C440 S68

An essay consisting of four main sections: (1) music and worship practices in the Puritan churches of the seventeenth century; (2) music reform in the churches during the early eighteenth century; (3) the rise of the singing school movement; and (4) the decline of the old style of psalm-singing. Facsimiles; classified bibliography of fourteen writings and five music scores.

479. Cooke, Nym. "American Psalmodists in Contact and Collaboration, 1770–1820." Ph.D. dissertation. Ann Arbor: University of Michigan, 1990. xliii, 620 p.

Organized into two parts: (1) personal contacts and collaborations and (2) musical contacts. Part 1: Examines the contacts and interactions between early American psalmodists, particularly in rural Massachusetts and Connecticut, exemplified by composers within musical families, within musical communities, composers who collaborated together in the music business, and within music associations. Part 2: Examines the influence composers had upon others, presenting examples of borrowing and modeling. Musical examples, facsimiles, and illustrations; alphabetical listing of all American psalmodists whose music was printed or appeared in printed or manuscript anthologies through 1810, including compilers; table of modeling, borrowings, and influence between composers; bibliography of approximately 180 writings and three hundred music manuscripts and early music imprints; expansive index to part 1.

480. Crawford, Richard. *The American Musical Landscape: The Business of Musicianship from Billings to Gershwin.* Berkeley: University of California Press, 2000. xix, 381 p. ISBN 0–520–22482–5 ML200 .C68 2000

Electronic version: Boulder, Colorado: NetLibrary, 2000. URL: http://www.netLibrary.com/urlapi.asp?action=summary&v=1&bookid=21182

Updated edition, with new preface, of Crawford's *The American Musical Landscape* (1993). One chapter titled "William Billings (1746–1800) and American Psalmody: A Study of Musical Dissemination." The majority of the book researches secular music; however, the index is helpful in locating other references to sacred music. Primarily concerned with economic history within the music industry. Musical examples and tables; general bibliography for the book of approximately 240 writings, fourteen editions and collections of music, five musical manuscripts, and six recordings; expansive index. Also published as an e-book. Mode of access: World Wide Web.

481. Crawford, Richard. "'Ancient Music' and the Europeanizing of American Psalmody, 1800–1810" in *A Celebration of American Music: Words and Music in Honor of H. Wiley Hitchcock,* ed. Richard Crawford, R. Allen Lott, and Carol J. Oja. Ann Arbor: University of Michigan Press, 1990. p. 225–255. ISBN 0–472–09400–9 ML200 .C44 1990

Discusses the viewpoints of various historians on European influences within early-nineteenth-century psalmody in the United States. Tables; bibliography of approximately seventy writings.

482. Haraszti, Zoltán. *The Enigma of the Bay Psalm Book.* Chicago: University of Chicago Press, 1956. xiii, 143 p. BS1440.B415 H3

Companion volume to *The Bay Psalm Book: A Facsimile Reprint of the First Edition of 1640* (University of Chicago Press, 1956). Discusses contributors to and content and printings of the New England psalm book. Of special note is Chapter 8, "The Psalm-Singing of the Puritans." Facsimiles and illustrations; documented with endnotes; index of names.

483. Hood, George. *A History of Music in New England: With Biographical Sketches of Reformers and Psalmists.* Boston, Mass.: Wilkins, Carter & Co., 1846. vii, 252 p. ML200 .H77; reprints (microforms), Ann Arbor, Mich.: University Microfilms International, 1956; Nashville, Tenn.: Historical Commission, Southern Baptist Convention, 1966; Chicago.: Library Resources, 1970 (rep. 1971); Louisville, Ky.: Lost Cause Press, 1975; Evanston, Ill.: American Theological Library Association, 1990;

reprint (book), New York: Johnson Reprint Corp., 1970. Intro. by Johannes Riedel. xxiii, vii, 252 p. ML200 .H77 1970

Originally published nearly 160 years ago. Reprinted a number of times. A study of psalmody in New England from its first settlement to 1800. Presents an annotated list of more than forty collections of music published between 1774 and 1799. Biographical sketches of seven reformers and ten psalmists. Documented with footnotes; index.

484. Inserra, Lorraine, and H. Wiley Hitchcock, eds. *The Music of Henry Ainsworth's Psalter (Amsterdam, 1612).* Brooklyn, N.Y.: Institute for Studies in American Music, 1981. vii, 126 p. ISBN 0–914678–15–9 M2117 .M976

Originated as Inserra's M.A. thesis (Brooklyn College of the City University of New York, 1980), with Hitchcock serving as adviser. Presents musical and poetical transcriptions of 39 tunes from the Ainsworth Psalter, brought to the Plymouth Colony by the Pilgrims. Includes critical commentary and historical information. Music, facsimiles, tables; bibliography of fourteen writings and discography of three recordings. Inserra's original thesis not available for perusal, but said to contain transcriptions of all 150 tunes from the Ainsworth Psalter.

485. Macdougall, H. C. *Early New England Psalmody: An Historical Appreciation, 1620–1820.* Brattleboro, Vt.: Stephen Daye Press, 1940. x, 179 p. ML200.3.M23 E2; reprint, New York: Da Capo Press, 1969. ML200.3.M23 E2 1969

Sacred music in New England from 1620 to 1820. Introductory chapters treat music of the Protestant Reformation in Bohemia, Germany, France, England, and Scotland. Following chapters discuss music in America. Topics discussed: Reformation psalters; influence of Thomas Ravenscroft, John Playford, and Tate & Brady on New England psalmody; *The Bay Psalm Book* (1640); William Billings; fuguing tunes; and singing schools. Musical examples, facsimiles, and tables; documented with endnotes; expansive index.

486. Meeter, Daniel J. "Genevan Jigsaw: The Tunes of the New-York Psalmbook of 1767" in *Ars et Musica in Liturgia: Celebratory Volume Presented to Casper Honders on the Occasion of His Seventieth Birthday on 6 June 1993,* ed. Frans Brouwer and Robin A. Leaver. Utrecht: Nederlands Instituut voor Kerkmuziek, 1993. pp. 150–166. ML3000 .A74

Meeter, Daniel J. "Genevan Jigsaw: The Tunes of the New-York Psalmbook of 1767" in *Ars et Musica in Liturgia: Essays Presented to Casper*

Honders on His Seventieth Birthday, ed. Frans Brouwer and Robin A. Leaver. Metuchen, N.J.: Scarecrow Press, 1994. pp. 150–166. ISBN 0–8108–2948–7 ML3000 .A74 1994

Relationship between the original Genevan psalter and a psalmbook commissioned by the Dutch Reformed Church in New York. Tables; documented with footnotes; no index.

487. Moore-Kochlachs, Emma Caroline. "The Psalms in the Worship of the Church Today." D.Min. dissertation. California: School of Theology at Claremont, 1990. vi, 187 p.

Use of psalms in the Jewish and Christian church traditions. Focuses on the ancient Jewish Temple and present-day synagogue and the early Christian church, the medieval monastic community, and twentieth-century Protestantism as represented by American Methodist traditions. Musical examples; glossary of terms; bibliography of nearly two hundred writings and music resources (books, periodicals, psalm translations and paraphrases, metrical psalters, chant psalters, hymnals and worship books, liturgical resources, and organ solos based on psalm tunes and psalm tones).

488. Pratt, Waldo Selden. *The Music of the Pilgrims: A Description of the Psalm-Book Brought to Plymouth in 1620.* Boston, Mass.: Oliver Ditson; Chicago: Lyon & Healy; New York: Chas. H. Ditson, 1921. 80 p. ML200.2 .P7; reprint, New York: Russell & Russell, 1971. ML200.2 .P7 1971

Very dated. Pratt's study of Henry Ainsworth's Psalter, to which the title refers, presents translations of the prose with critical commentary, fashions metrical arrangements of each, and provides a transcription in modern notation of thirty-nine melodies from the Psalter. It would be wise to compare Pratt's study with more recent research, including *The Music of Henry Ainsworth's Psalter (Amsterdam, 1612)* edited by Lorraine Inserra and H. Wiley Hitchcock (item 484). Facsimile of the Ainsworth's Psalter title page.

489. Worst, John William. "New England Psalmody, 1760–1810: Analysis of an American Idiom." Ph.D. dissertation. Ann Arbor: University of Michigan, 1974. xxi, 543 p.

Two parts. Part I discusses the decline of psalmody and the rise of hymnody in New England over a seventy-year period with description of stylistic characteristics of the music: scale, key, tonality, and harmony; rhythm, meter, and declamation; discord and harmony; melody; and texture. Part II presents music in full notation the 101 psalm and hymn tunes most frequently printed before 1810, along with descriptive tables. Numerous musical examples and tables; bibliography of more than one hundred writings.

REQUIEMS

490. Clency, Cleveland Charles. "European Classical Influences in Modern Choral Settings of the African-American Spiritual: A Doctoral Essay." D.M.A. dissertation. Coral Gables, Fla.: University of Miami, 1999. ix, 154 p.

Study of the influence of the European tradition on the spiritual in the United States. Biographical information provided for fifteen composers/arrangers. In-depth analysis of a dozen spirituals. Musical examples; annotated bibliography of more than one hundred writings.

SPIRITUALS

491. Cone, James H. *The Spirituals and the Blues: An Interpretation.* New York: Seabury Press, 1972. viii, 152 p. ISBN 0–8164–0236–1, ISBN 0–8164–2073–4 (pbk.) ML3556 .C66; reprint, Westport, Conn.: Greenwood Press, 1980. ISBN 0–313–22667–9 ML3556 .C66 1980; reprint, Maryknoll, N.Y.: Orbis Books, 1991 (1992). ix, 141 p. ISBN 0–88344–747–9, ISBN 0–88344–843–2 (pbk.) ML3556 .C66 1991

First published in 1972 and reprinted several times. Examines the African-American spiritual and the blues, described as a secular spiritual, within a social, historical, and religious perspective. Documented with endnotes.

492. Dixon, Christa K. *Negro Spirituals: From Bible to Folk Song.* Philadelphia: Fortress Press, 1976. x, 117 p. ISBN 0–8006–1221–3 ML3556 .D58

Relates twenty-three popular spirituals to biblical scripture. Selected spirituals discussed include: "Swing Low, Sweet Chariot," "Go Down, Moses," "Joshua Fit de Battle ob Jericho," "Sometimes I Feel Like a Motherless Chile," "Go Tell It on the Mountain," "Were You There When They Crucified My Lord?," "When De Saints Come Marchin' In," and "He's Got the Whole World in His Hands." Index to spiritual references and general index.

493. Lovell, John, Jr. *Black Song: The Forge and the Flame: The Story of How the Afro-American Spiritual Was Hammered Out.* New York: Macmillan, 1972. xviii, 686 p. ML3556 .L69; reprint, New York: Schirmer Books, 1980. ISBN 0–02–871900-X ML3556 .L69 1980; reprint, New York: Paragon House, 1986. ISBN 0–913729–53–1 ML3556 .L69 1986

Surveys the origin and development of the African-American spiritual within a cultural perspective. Quite lengthy. Reviews approximately five hundred spirituals. Illustrations, photos, and charts; bibliography of about 250 writings; index to spiritual titles and expansive general index.

See also: Collins, Mary, David Power, and Mellonee Burnim, eds. *Music and the Experience of God* (item 629); Jackson, George Pullen. *White Spirituals in the Southern Uplands: The Story of the Fasola Folk, Their Songs, Singings, and "Buckwheat Notes"* (item 406); Kaatrud, Paul Gaarder. "Revivalism and the Popular Spiritual Song in Mid-Nineteenth Century America: 1830–1870" (item 456); Rogal, Samuel J. *A General Introduction to Hymnody and Congregational Song* (item 443)

VOCAL MUSIC

494. Lawhon, Sharon L. "A Performer's Guide to Selected Twentieth-Century Sacred Solo Art Songs Composed by Women from the United States of America." D.M.A. dissertation. Louisville, Ky.: Southern Baptist Theological Seminary, 1993. xiii, 101 p.

Study of twentieth-century sacred solo art songs by U.S. women composers. Analyzes selected works by H. H. A. (Amy) Beach, Emma Lou Diemer, Clara Edwards, Carrie Jacobs-Bond, Mana-Zucca, Julia Perry, Lily Strickland, and Elinor Remick Warren. Musical examples; a listing of sacred solo art songs composed by eighty-two U.S. women composers; title, biblical reference, and topical and functional indexes; bibliography of sixty-six writings.

VIII

Music Ministry

GENERAL WORKS

495. Ball, Louis, and Mary Charlotte Ball, eds. *On the State of Church Music, V: Symposium on Church Music in the Twenty-First Century: Papers.* Jefferson City, Tenn.: Louis and Mary Charlotte Ball Institute of Church Music, Center for Church Music, Carson-Newman College, and Tennessee Baptist Convention, Church Music, 1997. 68 p. ML3106 .O5 1997

Fourteen diverse essays on Protestant church music in the United States. Essays: "The Key to Church Growth" (M. C. Alford); "New Ways: Responsive and Responsible Church Ministry Leadership Approaching the 21st Century" (W. J. Boertje); "Gather into One: Praying and Singing Globally" (C. M. Hawn); "Spotlights, Dancers, and Hymns" (R. W. Hicks); "Guidelines for Music Ministry in Large Presbyterian (U.S.A.) Churches" (R. G. Miller); "Where Will All the Musicians Come From? Music Ministry Skills for the 21st Century" (G. M. Pysh); "Sacred Music at the Cinema: Jacques Ibert and Golgotha" (W. M. Roberts); "Church Music in the 21st Century: A Symposium" (T. Sharp); "Church Music in the Vernacular" (W. D. Simpson); "Reinvent or Recover?: Considering Worship for a New Millennium" (R. A. Smith); "Virtual Worship" (Julian S. Suggs); "Tastes Great, Less Filling" (Tracy Wilson); "New Trends in 21st Century Music and Worship" (Y. S. Wong); and "The Graded Choir as Catalyst to Worship Participation" (G. G. Wyrick). Bibliographies follow some of the essays.

496. Bish, Diane. *Church Music Explosion: A Practical Challenge for the Church Musician (With Aids for Ministers).* Foreword by D. James Kennedy. Ft. Lauderdale, Fla.: Joy Productions, 1977. 76 p. ML3100 .B3; reprint, S.l.: Fred Bock Music, 1982. 78 p. ML3100 .B3 1982

Advice for the church musician. Covers directing a church choir, the church organist, and congregational singing. Musical examples, illustrations, and photos; bibliography of about 180 choral works and a classified bibliography of approximately 240 organ works.

497. Burroughs, Bob. *An ABC Primer for Church Musicians.* Foreword by James D. Woodward. Nashville, Tenn.: Broadman Press, 1990. 144 p. ISBN 0–8054–3307–4 ML3001 .B93 1990

Twenty-six chapters, each title beginning with successive letters of the alphabet, A through Z (Alto; Basses; Committees, etc.). Offers suggestions to music directors and church musicians. Often humorous reading.

498. Bye, L. Dean. *God's People Worship!: A Manual for Building Powerful Worship.* Pacific, Mo.: Mel Bay Publications, 1984. 212 p. ISBN 0–87166–007–5 MT88 .B83

Manual for the church musician. Treats melody, rhythm, harmony, expression, and use of instruments and dance. Four appendixes: (1) music signs and expression marks, (2) definitions of common musical terms, (3) glossary, and (4) guitar chords. Numerous illustrations and musical examples; index of song titles.

499. Causey, C. Harry. *Open the Doors . . . : A Motivational Guide and Practical Aid for Creativity in Today's Corporate Worship Services.* 2nd ed. Introductions by Jerry R. Kirk and Richard C. Halverson. Rockville, Md.: Music Revelation, 1985. 114 p. MT88 .C38 1985

Based on the author's experiences as a church musician. Organized into fifty brief chapters, covering recruitment of musicians, publishing a bulletin, decorative ideas, selection of music, and many other topics relating to music within corporate worship. Brief bibliographies of music for some chapters.

500. *Church Music Handbook.* Paul Hamill, M. Elinor Hamill, eds. Great Barrington, Mass.: Gemini Press International, 1982–. ISBN 0–9679778–4–3 (2003 ed.) ML3111 .H16

Published annually since 1982. "Designed for use by clergy and church musicians in the Presbyterian Churches, the United Church of Christ and the United Methodist Church." The largest portion of each issue, approximately two-thirds of the text, is devoted to the church year with a list of recommended music (hymns, anthems, and organ music) and scripture readings for each Sunday and holy day. Other features include a few articles, sample music, and bibliographies of newly published choral, handbell, and organ music. Index of first lines and hymn names.

501. Collins, Donnie Lee. "Principles and Practices Prevailing in Church Music Education Programs of Selected Protestant Churches of America." Ph.D. dissertation. Tallahassee: Florida State University, 1970. vii, 294 p.

Purpose: "to study principles and practices prevailing in music education programs in selected churches of the Methodist, Baptist, Lutheran, Episcopal, and Presbyterian denominations of America." Includes brief reviews of church music education and denominational philosophies and objectives. Develops guidelines for new and emerging programs. Numerous appendixes and a bibliography of approximately 150 writings.

502. Corbitt, J. Nathan. *The Sound of the Harvest: Music's Mission in Church and Culture.* Grand Rapids, Mich.: Baker Books; Carlisle, Cumbria, United Kingdom: Solway, 1998. 352 p. ISBN 0–8010–5829–5 (U.S.), ISBN 1–900507–88–9 (U. K.) ML3001 .C785 1998

Part 1: Addresses the purpose of music within church communities, exploring the role of music as priest, prophet, proclaimer, healer, preacher, and teacher. Part 2: Offers biblical principles for evaluating music, discussing primary forms of music in Christian churches, surveying instruments commonly used in worship, and addressing ways to achieve effective music leadership. Illustrations; bibliography of about 170 writings; expansive index.

503. Delamont, Vic. *The Ministry of Music in the Church.* Chicago.: Moody Press, 1980. 160 p. ISBN 0–8024–5673–1 ML3110 .D44

Practical guide to music ministry directed toward "that yet multitudinous group of churches and peoples known as the typical North American evangelical church, be it Baptist, Alliance, Evangelical Free, Nazarene, or other." Topics include developing a philosophy of music ministry, qualifications of music personnel, organization and administration of the music program, music groups, determining appropriateness, rehearsals, special productions, music and Christian education, and music and worship. Illustrations; classified bibliography of sixty-four writings; index.

504. Frink, George M. D. *The Music Director's Necessary Book.* 2nd ed. Charleston, S.C.: Carol Press, 2000. 126 p. ISBN 0–9679882–4–1 MT88 .F75 2000

Helpful suggestions for running a successful church music program. Covers office procedures, budget, organizing a music library, copyright issues, publicity, organizing a choir, and preparing music for worship. Illustrations and charts.

505. Hunter, William C. *Music in Your Church: A Guide for Pastors, Music Committees, and Music Leaders.* Valley Forge, Pa.: Judson Press, 1981. 109 p. ISBN 0–8170–0917–5 ML3001 .H88

Addresses responsibility for church music in regard to pastor, music committees, choir director, and choir members. Also discusses youth and children's choirs, with a few tips for music instruction. Illustrations.

506. Lawrence, Joy E., and John A. Ferguson. *A Musician's Guide to Church Music.* Foreword by Paul Manz. New York: Pilgrim Press, 1981. xiii, 255 p. ISBN 0–8298–0424–2 MT88 .L39

A variety of topics suitable for the church musician, including selecting music, working with adult and children's choirs, use of instruments, use of interpretive dance, and designing, purchasing, and maintaining the organ. Musical examples, tables, photographs, and illustrations; repertory lists, a few works by U.S. composers, of approximately 165 vocal solos, fifteen duets, ninety works for organ, and forty works for recorder; classified bibliography (choral; church history; church music; hymnology; organ; general) of more than two hundred writings; expansive index.

507. Levy, Ezekiel. "Sacred Music and the Festivals of the Catholic, Jewish, and Protestant Faiths." Ed.D. dissertation. New York: New York University, 1955. ix, 299 p.

Purpose: "to select representative songs from the sacred music of Catholic, Jewish, and Protestant religious festivals and holidays, to relate each song to its contemporary and historic religious and secular settings, and to suggest ways in which these materials may be used to promote mutual respect and understanding among persons who differ in religious backgrounds." Musical examples and tables; bibliography of more than one hundred writings.

508. Lewis, Samuel T. *Facing the Music in the Local Church.* Preface by George E. Tutwiler. Tyrone, Pa.: Kathryn H. Lewis, 1998. vii, 39 p. MT88 .L48 1998

Practical suggestions for improving the quality of music in worship, aimed toward choir directors, organists, and worship leaders. Author draws from his experiences as a music minister and pastor in the Presbyterian Church. Musical examples; bibliography of nine recommended readings.

509. Lovelace, Austin C., and William C. Rice. *Music and Worship in the Church.* Rev. and enl. Nashville, Tenn.: Abingdon Press, 1976 (rep. 1987).

256 p. ISBN 0–687–27358–7 (1976), ISBN 0–687–27357–9 (1987) ML3100 .L7 1976, ML3100 .L7 1987

Focuses on Protestant church music. Separate chapters discuss the music committee, the director, the organist, the adult choir, children's and youth choirs, the choir's music, the soloist, the congregation, music in Christian education, and contemporary music and worship. Many chapters conclude with annotated lists of recommended music. Glossary of terms; classified bibliography of more than 350 writings, music collections, and periodicals; expansive index.

510. Mitchell, Robert H. *Ministry and Music.* Philadelphia: Westminster Press, 1978. 163 p. ISBN 0–664–24186–7 ML3100 .M55

Practical book regarding congregational singing, organ playing, and organizing and directing a choir. But book's higher purpose is to examine "how a minister can relate his theology to the goals and contributions of music and how the church musician can relate his musical expertise to the broader concept of Christian ministry." Discusses debate over inclusion of popular music in worship services. Documented with endnotes.

511. Orr, N. Lee. *The Church Music Handbook: For Pastors and Musicians.* Nashville, Tenn.: Abingdon Press, 1991. 144 p. ISBN 0–687–01624X MT88 .O77 1991

Team-building approach to issues confronting church music at the close of the twentieth century. List of fourteen organizations and their publications for church musicians; bibliography of twenty-one writings; no index.

512. Owens, Bill. *The Magnetic Music Ministry: Ten Productive Goals.* Edited by Herb Miller. Nashville, Tenn.: Abingdon Press, 1996. 109 p. ISBN 0–687–00731–3 MT88 .O95 1996

Concerns improvement of the church music program. Suggestions range from budget concerns to establishing leaders and support groups to how to develop effective rehearsals. Documented with endnotes.

513. Routley, Erik. *Church Music and the Christian Faith.* Foreword by Martin E. Marty. Carol Stream, Ill.: Agape, 1978. vi, 153 p. ISBN 0–916642–10–0 ML3001 .R83; reprint, London: Collins, 1980. 156 p. ISBN 0–00–599650–3 ML3001 .R83 1980

Revision of earlier title, *Church Music and Theology* (Fortress Press, Muhlenberg Press, SCM Press, 1959). Routley shares his insights on musical aesthetics, practical matters for choirs, directors, and organists, and theological issues. Not exclusively devoted to music of the United States; significant inclusion of European church music. Musical examples; bibliography of twenty-eight writings and annotated bibliography of five varied organ accompaniments, along with fragments and examples; expansive index.

514. Routley, Erik. *The Divine Formula: A Book for Worshipers, Preachers and Musicians and All Who Celebrate the Mysteries.* Foreword by Daniel Jenkins. Princeton, N.J.: Prestige Publications, 1986. iv, 166 p. ISBN 0–911009–03–5 BV176 .R68 1986

Routley shares his insights on preaching, liturgy, prayer, and the place of music in the worship service. Generally, only a paragraph is written about each subject. The subject index is useful for locating information on music. Subjects to consider: America, church music in; anthems; Baptist hymnal; cathedral music; choral music; English church music; gospel songs; hymns; musical and unmusical; organists; and pianoforte.

515. Routley, Erik. *Music Leadership in the Church: A Conversation Chiefly with My American Friends.* Nashville, Tenn.: Abingdon Press, 1967. 127 p. ML3000 .R69; reprint, Carol Stream, Ill.: Agape, 1984. 136 p. ISBN 0–916642–24–0 ML3000 .R69 1984

Comparison of musical practices in British and American churches. Three sections: (1) "The Church Musician and History," (2) "The Church Musician and His Bible," and (3) "The Church Musician and Worship." Musical examples; documented with endnotes; index of scripture references; index of names and subjects.

516. Smith, Willie Eva. *O Sing unto the Lord a New Song: A Complete Handbook for the Total Church Music Program.* New York: Vantage Press, 1976. 14, 73 p. SBN 533–01936–2 MT88 .S6

Practical advice concerning the church music program. Addresses the roles and relationships of the pastor, the director, the accompanists, and the choir.

517. Steere, Dwight. *Music in Protestant Worship.* Richmond, Va.: John Knox Press, 1960. 256 p. ML3100 .S72

Three main sections: (1) description of physical aspects of the church building; (2) the minister and church musicians; and (3) the music, including the hymn, processional and recessional, congregational service music (Gloria Patri, doxology, offertory response, and benediction response), anthem, choral response, organ music, solo song, and occasional service music (communion, funeral service, and wedding). Musical examples; classified bibliography of more than fifty writings; expansive index.

518. Terry, Lindsay. *How to Build an Evangelistic Church Music Program.* Foreword by Elmer L. Towns. Nashville, Tenn.: Thomas Nelson, 1974. ix, 198 p. ISBN 0–8407–5581–3 MT88.T465 H6

Based on the author's personal experience in building church music programs. Addresses such topics as organizing choirs, rehearsals, preparing music for services, television and radio, revivals and conferences, and other helpful suggestions. Musical examples, photographs, and conducting patterns.

519. Thayer, Lynn W. *The Church Music Handbook: A Handbook of Practical Procedures and Suggestions.* Grand Rapids, Mich.: Zondervan Publishing House, 1971. 190 p. MT88 .T5

Designed for pastors, church musicians and music committees, and congregations in Protestant churches. Discusses recommended qualifications for music directors/organists. Offers advice concerning choir rehearsals, use of musical instruments, congregational singing, special services and programs, and other practical matters. Illustrations; classified bibliography (general church music; Christian education; general church history; worship; adult choirs; age-group choirs; hymnology; organ, piano; church music history; liturgy, chants, responses; music buildings, rooms, equipment; specific, occasional, special; training and techniques; signs, symbols, visuals) of approximately 120 writings; expansive index.

520. Topp, Dale. *Music in the Christian Community: Claiming Musical Power for Service and Worship.* Grand Rapids, Mich.: W. B. Eerdmans, 1976. 205 p. ISBN 0–8028–1642–8 ML3111 .T66; reprint, Ann Arbor, Mich.: University Microfilms International, 1985. ML3111 .T66 1985

Protestant church music in the United States Addresses music in the Christian church, Christian home, and Christian school. Musical examples and tables; classified discography of one thousand recordings; no bibliography; no index.

521. Tredinnick, Noel, ed. *On the State of Church Music, II.* Jefferson City, Tenn.: Louis and Mary Charlotte Ball Institute of Church Music, and Center for Church Music, Carson-Newman College, 1994. 32 p. ML3106 .O5 1994

Three essays on current church music issues. Essays: "Qualifications for a Minister of Music" (Tredinnick); "Church Music: Status, Curriculum, Reaction" (L. O. Ball); and "Revolution in Worship and Church Music" (D. P. Hustad). The Tredinnick essay concerns English church music, but information relates to U.S. church music leadership as well.

522. Walters, Chéri. *Advice to the Minister of Music: Get a Giant Hat Rack!* Springfield, Mo.: Chrism, 1994. 142 p. ISBN 0–88243–339–3 ML3001 .W35 1994

Practical advice for the many roles of a minister of music: as a minister, staff member, counselor, communicator, manager, talent agent, producer, sound technician, and musician. Illustrations and tables; classified bibliography (children and youth; church music ministry; devotionals; hymnody; music and worship; musicianship: conducting, vocal, and instrumental techniques; organization and time management; other books of interest to music ministers; relationships; sound, lighting, and drama; women in leadership) of approximately 180 writings and various audio and video resources, computer software, conferences and worships, directories and catalogs, and addresses for musical rental and resale. Documented with endnotes.

523. Westermeyer, Paul. *The Church Musician.* Rev. ed. Minneapolis, Minn.: Augsburg Fortress, 1997. xvi, 159 p. ISBN 0–8066–3399–9 ML3001 .W5 1997

Practical suggestions for music leaders and musicians in Christian churches. Annotated bibliography of approximately one hundred writings; expansive index.

524. Whaley, Vernon M. *The Dynamics of Corporate Worship.* Grand Rapids, Mich.: Baker Books, 2001. 191 p. ISBN 0–8010–9109–8 BV15 .W44 2001

Chapters 7 and 8 are devoted to the integration of worship and music, the purpose of music in worship, the preservation of cultural and musical heritage, and biblical teachings concerning the use of music. Documented with endnotes; general bibliography of seventeen writings; no index.

525. Whaley, Vernon M. *Understanding Music & Worship in the Local Church.* Wheaton, Ill.: Evangelical Training Association, 1995. 96 p. ISBN 0–910566–65–8 BV290 .W43 1995

Chapter 7 presents a brief overview of evangelical church music in the United States from 1800 to the 1970s. Following chapters promote renewed emphasis on music and worship and on building an effective church music program. Chapter 12, "Resources for Music & Worship," provides addresses and phone numbers for a few hymnal publishers, choral clubs, sources of printed music, and other useful sources. Glossary of terms; bibliography of twenty-nine writings.

RELIGIOUS AND ETHNIC GROUPS

African American

526. Abbington, James. *Let Mt. Zion Rejoice!: Music in the African American Church.* Valley Forge, Pa.: Judson Press, 2001. xvi, 143 p. ISBN 0–8170–1399–7 ML3111 .A23 2001

Current state of music in the African-American church. Observes eight categories of musicians serving in churches today, addresses fundamentals for church musicians, and considers the relationship between pastors and musicians. Discusses choirs, the importance of planning worship, and closely examines three genres of music: hymns, anthems, and congregational spirituals. Bibliography of approximately 130 writings; no index.

527. Harris, James H. "How Shall We Preach without a Song: A Model for African American Sacred Music." D.Min. dissertation. Dayton, Ohio: United Theological Seminary, 1996. 67, xvii p.

Addresses "the need for relevant music in worship expressing Black theological themes as presented in African American preaching." Study conducted at Brown Chapel African Methodist Episcopal Church, Cincinnati, Ohio. Statistical information; bibliography of fourteen writings.

528. Mapson, J. Wendell. *The Ministry of Music in the Black Church.* Valley Forge, Pa.: Judson Press, 1994. 96 p. ISBN 0–8170–1057–2 MT88 .M17 1984

Purpose: "to provide guidelines for the qualitative use of music in the black church." Addresses music and worship in African-American culture, responsibilities of the pastor and church musicians, and music in various contexts of worship. Bibliography of about eighty writings; no index.

529. Mapson, J. Wendell. "Some Guidelines for the Use of Music in the Black Church." D.Min. dissertation. Philadelphia: Eastern Baptist Theological Seminary, 1983. viii, 169 p.

Purpose: "to help contemporary Black ministers to become more aware of issues related to ministry and music by examining the use of music in the Old and New Testament, as well as surveying the historical and cultural context out of which music and worship in the Black church has evolved." Some data obtained through questionnaires administered to ministers and musicians; results tallied into tables. Bibliography of eighty-one writings.

Apostolic Faith Churches of God

530. Noble, E. Myron. *The Gospel of Music: A Key to Understanding a Major Chord of Ministry.* Foreword by John L. Meares. Introduction by Pearl Williams-Jones. Washington, D.C.: Middle Atlantic Regional Press of the Apostolic Faith Churches of God, 1986. xxvi, 159 p. ISBN 0–9616056–1–8 ML2900 .N62 1986

Examines scriptural references on the use of music in worship. Three sections: "Music Geneses," "Performance and Support Modes," and

"Practicum and Curriculum." Section 1: pre-Christian overview; Judeo-Christian transition; textual content; categories of songs; New Testament nomenclature; and musical and non-musical instruments. Section 2: authority, qualification, and training of performers of music; men and/or women; forms of performance; performance dynamics (life cycle of man; musical; liturgical; pastoral; personal hygiene, personal attire); world performance; and remunerations (salary or free will). Section 3: Christian music leadership; models for administration; educating the household of faith; music for congregation; music stewards; and seminary music curricula. Includes seven sermon outlines and about fifty titles with biblical references. Bibliography of nine writings.

Baptist

531. Anderson, William M. *And a Music Director, Too.* Updated ed. Nashville, Tenn.: Convention Press, 1990. 80 p. ML3869 .A64 1990

Text for a course in church music leadership, sponsored by The Sunday School Board of the Southern Baptist Convention. Illustrations; classified bibliography of more than one hundred writings.

532. Armstrong, Gerald P., Jere V. Adams, and Clinton E. Flowers, compilers. *The Music Ministry Resource Manual: For Creative Church Musicians.* Edited by Adams. Nashville, Tenn.: Convention Press, 1990. Various paginations. ML3160 .A76 1990

Text for a course in the Church Study Course of the Baptist Sunday School Board. Variety of topics relating to Protestant church music ministry with multi-sectional chapters written by various authors. Contributors are: G. C. Adkins, W. M. Anderson, Jr., D. Baker, N. J. Blair, D. Blakley, P. Bobbitt, L. E. Brooks, K. Butler, H. E. Cone, P. Duke, M. Edwards, R. Edwards, T. Evans, M. Ezell, W. L. Forbis, Bill Green, B. J. Green, B. Hatfield, J. R. Hawkins, J. Helman, R. Hewell, M. Hodges, T. Holcomb, R. Jackson, D. Johnson, F. L. Kelly, T. Keown, A. J. King, M. H. Kirkland, R. Lambros, J. L. Landrum, M. Lawson, J. Lee, J. Lewis, R. Lewis, T. Lyons, J. McCaleb, D. L. McClard, G. E. McCormick, D. G. McCoury, B. Nehring, M. K. Parrish, B. Pearson, E. Porter, J. Raymick, P. Riddle, E. Robertson, B. Roper, B. Sanders, C. E. Sherman, M. Short, Jr., H. Smeltzer, L. G. Smith, A. M. Smoak, Jr., E. Spann, T. M. Stoker, R. Stone, H. Taylor, H. Thomas, H. L. Webb, Jr., I. West, H. Wicker, M. Wilkins, T. Willoughby, and T. W. York. Illustrations.

533. Bearden, Donald Roland. "Competencies for a Minister of Music in a Southern Baptist Church: Implications for Curriculum Development." Ph.D. dissertation. Baton Rouge: Louisiana State University, 1980. x, 219 p.

Examines the skills, behaviors, and knowledge a successful minister of music in a Southern Baptist church should possess. Topic introduced with background information on church music and Southern Baptists. Addresses hymnody, worship planning, musicianship and musical performance, choral conducting, children's music, instrumental music, and church music administration. Tables; bibliography of nearly eighty writings.

534. Bobbitt, Paul, and Gerald P. Armstrong. *The Care and Feeding of Youth Choirs.* Nashville, Tenn.: Convention Press, 1975. 112 p. MT88 .B6

Designed as a text for a course in church music sponsored by The Sunday School Board of the Southern Baptist Convention. A practical book on building and maintaining a youth choir program. Discusses basics such as motivation, selecting music, rehearsals, and ministering to youth. Musical examples and illustrations.

535. Easterling, R. B. *Church Music for Youth.* Nashville, Tenn.: Convention Press, 1969. v, 122 p. ML3160 .E28

Textbook developed by The Sunday School Board of the Southern Baptist Convention, which "presents and explains the philosophy, concepts, and organizational patterns of the Youth Division of the Church Music program." Charts and tables; bibliography of approximately fifty writings.

536. Ham, Richard. *Church Music for Children.* Nashville, Tenn.: Convention Press, 1969. vii, 113 p. ML3160 .H3

Textbook developed by The Sunday School Board of the Southern Baptist Convention. Practical instruction for organizing and developing a church music program for children, including discussion of ministry through music, space considerations, needed equipment and supplies, and financial planning. Charts; bibliography of approximately 150 scores and music collections, seventy writings, and a few recordings for children.

537. Keown, Tommy, Derrell Billingsley, Martha Kirkland, and William M. Anderson, Jr., compilers. *Growing a Musical Church: A Handbook for Music Leaders in the Small Church.* Edited by William M. Anderson, Jr. Nashville, Tenn.: Convention Press, 1990. 128 p. ML3160 .G76 1990

Designed as text for course in the Church Study Course of the Baptist Sunday School Board. Handbook of fourteen chapters, each chapter or portion of chapter written by various authors. Aimed toward small churches, subjects include music for congregation, instrumental music (keyboard accompaniment, electronic music, band, and handbells), adult choir, children and youth choirs, music for soloists and ensembles, and other topics relating to budget, materials, space concerns, and outreach. Contributors are: Keown, Kirkland, Anderson, G. K. Hitt, A. L. "Pete" and D. J. Butler, P. Riddle, B. Orton, M. W. Lawson, N. Grantham, D. R. Jones, J. McCaleb, J. and S. Cauley, L. Konig, and M. Ezell. Musical examples, illustrations, and photos.

538. Reynolds, William Jensen. *Building an Effective Music Ministry.* Nashville, Tenn.: Convention Press, 1980. 143 p. MT88. B85 1980

Text for a course in church music leadership, sponsored by The Sunday School Board, Woman's Missionary Union, and Brotherhood Commission of the Southern Baptist Convention. Discusses administrative techniques and effective use of resources (time, budget, facilities, etc.).

539. Reynolds, William Jensen. *Congregational Singing.* Nashville, Tenn.: Convention Press, 1975. 119 p. ML3111 .R49

Nine brief chapters on planning, directing, and accompanying congregational singing, specifically associated with *Baptist Hymnal* (1975). Musical examples; bibliography of fifteen free organ accompaniments; suggested tempos for each hymn; index of hymns organized by key.

540. Suggs, Julian S. "The Program of Church Music Leadership Training for the Volunteer/Part-Time Director as Administered by State Music Departments of the Southern Baptist Convention." D.M.A. dissertation. Louisville, Ky.: Southern Baptist Theological Seminary, 1979. viii, 196 p.

Examines "the role of the state music director in providing church music leadership training for volunteer and part-time directors." Concludes that approximately 40 percent of Southern Baptist churches are served by volunteer music directors and that over one-half have not had formal training in music. Provides historical background on singing schools, shaped note singing, influences on Baptist church music, the influence of gospel hymnody, and the Southern Baptist training program. Tables; bibliography of nearly eighty writings; various appendixes, including a listing of state music departments.

541. Vaught, W. Lyndel. *Senior Adult Choir Ministry.* Foreword by James R. Hawkins. Nashville, Tenn.: Convention Press, 1991. x, 109 p. MT88 .V36 1991

Designed as a text for a course in church music sponsored by The Sunday School Board of the Southern Baptist Convention. Practical advice for directors of senior adult choirs. Deals with vocal technique and rehearsals, as well as ministry to choir members. Musical examples, illustrations, and photos; documented with footnotes with list of recommended readings at end of each chapter.

Catholic

542. Bauman, William A. *The Ministry of Music: A Guide for the Practicing Church Musician.* 2nd ed. Rev. by Elaine Rendler. Ed. by Thomas Fuller. Washington, D.C.: Liturgical Conference, 1979. iv, 124 p. ML3002 .B28 1979

Practical advice for the church musician in Catholic worship. Addresses music ministry, the cantor, choral groups, the organist, guitar and other instruments, repertoire, planning worship, liturgical and pastoral judgment, and special celebrations. Musical examples and illustrations.

543. Brownstead, Frank, and Pat McCollam. *The Volunteer Choir.* Washington, D.C.: Pastoral Press, 1987. vii, 72 p. ISBN 0–912405–37–6 MT88 .B9 1987

Practical suggestions for directors of volunteer choirs in Catholic churches. Includes classified, concisely annotated bibliography of approximately 170 recommended choral works, providing composer, title, publisher, and level of difficulty. Musical examples and illustrations; bibliography of twelve writings.

544. Connolly, Michael. *The Parish Cantor: Helping Catholics Pray in Song.* Rev. ed. Chicago.: G. I. A. Publications, 1991. 67 p. ISBN 0–941050–24–6 MT88 .C65 1991

Originally published in 1981, revised in 1991. Resource for parish cantor programs. Covers the cantor's responsibilities, qualifications, and skills, the Sunday Mass, musical forms and repertoire, helping people sing, and the parish cantor program. Musical examples. Three appendixes: (1) instructions for cantors from official documents; (2) bibliography of a few writings on cantor training and church music in general and music scores, as well as recommended publishers; and (3) outline of training exercises.

545. Eddy, Corbin T. *Musicians at Liturgy.* Minneapolis, Minn.: Winston Press, 1983. 38 p. ISBN 0–86683–737-X ML3795 .E33 1983

Booklet addressing Catholic church music issues. Discusses development of music programs, music committees, structure of the liturgy and music elements involved, and personal development of music program participants. Photographs; bibliography of twelve writings.

546. Haas, David. *The Ministry and Mission of Sung Prayer.* Cincinnati, Ohio: St. Anthony Messenger Press, 2002. x, 110 p. ISBN 0–86716–214–7 ML3002 .H26 2002

The author shares his experience in the field of music ministry. Ten chapters: "Reflections on Liturgy and Ritual," "Biblical Sources for Music Ministry," "Toward a Spirituality of Music Ministry," "Who Are the Music Ministers?," "Images for the Minister of Music," "What Kind of Music Do We Need?," "The Liturgy and Its Elements," "The Liturgical Year," "Preparing for the Liturgy: Resources and Process," and "Issues and Challenges: Present and Future." Tables; expansive index.

547. Haas, David. *Music & the Mass: A Practical Guide for Ministers of Music.* Chicago.: Liturgy Training Publications, 1998. xi, 129 p. ISBN 1–56854–198–8 ML3001 .H14 1998

Described by author as "a guidebook whose goal is to point musicians and liturgical ministers toward key documents and principles of celebration of the Eucharist." Each of four chapters discusses a main element of the Mass. Chapters: "The Gathering Rites," "The Liturgy of the Word," "The Liturgy of the Eucharist," and "The Concluding Rite." Documented with endnotes.

548. Hansen, James, Melanie Coddington, and Joe Simmons. *Cantor Basics.* Rev. ed. Portland, Ore.: Pastoral Press, 2002. xii, 160 p. ISBN 1–56929–042–3 MT88 .H29 2002

Instructional resource for cantors in the Catholic Church. Consists of 112 questions followed by answers. Organized into five subject areas: the assembly; the ministry of the cantor; the scriptures—the psalter; the responsorial psalm; and special people, special situations, special skills. Bibliography of fifty-four partially annotated recommended readings.

549. Hume, Paul. *Catholic Church Music.* Preface by Francis J. Guentner. New York: Dodd, Mead, 1956 (rep. 1957, 1960). xiv, 259 p. MT88 .H8

Addresses music ministry in the Catholic Church. Discusses the choir, boy's choir, congregational singing, the organist, music for weddings, and music in schools. Classified bibliography of a few recommended writings; classified listing of seventy recommended musical works (Masses and organ music); expansive index.

550. Johnson, Lawrence J. *The Mystery of Faith: The Ministers of Music.* Washington, D.C.: National Association of Pastoral Musicians, 1983. vi, 122 p. ISBN 0–9602378–9–5 ML3007 .J63 1983

Studies various aspects of Catholic music ministry. Considers assembly, presider, deacon, cantor, choir, instrumentalists, organists, dancer, and composer. Suggested questions for discussion and bibliographies following each topic. Illustrations; documented with endnotes.

551. Kodner, Diana. *Handbook for Cantors.* Rev. ed. Chicago: Liturgy Training Publications, 1997. xi, 116 p. ISBN 1–56854–097–3 MT860 .K65 1997

Handbook for cantors in Catholic churches. Glossary of terms; numerous musical examples; bibliographies of recommended reading follow some chapters.

552. Lovrien, Peggy. *The Liturgical Music Answer Book.* San Jose, Calif.: Resource Publications, 1999. 149 p. ISBN 0–89390–454–6 MT88 .L695 1999

101 questions with answers about Catholic church music grouped within the following subject areas: groundwork questions; the music ministry; basic instruments of the liturgy; implementing liturgical music; questions from choir leaders and members; questions from pastors and pastoral leaders; paying liturgical music directors; questions from the assembly; and unity within varying cultures. Annotated bibliography of twenty-six writings; expansive index.

553. McKenna, Edward J. *The Ministry of Musicians.* Collegeville, Minn.: Liturgical Press, 1983. 40 p. ISBN 0–8146–1295–4 ML3002 .M35 1983

Brief. Practical advice for church musicians in contemporary worship. Catholic slant.

554. Middlecamp, Ralph. *Introduction to Catholic Music Ministry.* Glendale, Ariz.: Pastoral Arts Associates of North America, 1978 (rep. with corrections, 1982). 47 p. ISBN 0–89699–012–5 ML3011 .M627

Five brief chapters: "Being and Becoming a Minister of Music," "Music in Eucharistic Worship," "Choosing Music for Celebration," "Some Practical Matters," and "Continuing to Grow in Your Ministry." Well organized. Much of the writing in outline form. Bibliography of nineteen writings.

555. Patterson, Keith L. *Evaluating Your Liturgical Music Ministry.* San Jose, Calif.: Resource Publications, 1993. viii, 140 p. ISBN 0–89390–258–6 MT88 .P35 1993

Provides analytical tools for evaluating liturgical music ministry, specifically within the Catholic Church. Organized into two parts: (1) "The Standards" and (2) "The Evaluation Model," which discusses the evaluation process. Bibliography of eleven writings.

556. Reid, Heather. *Preparing Music for Celebration.* Ottawa, Ontario, Canada: Novalis; Collegeville, Minn.: Liturgical Press; Alexandria, Australia: E. J. Dwyer, 1996. 47 p. ISBN 2–89088–801–0 (Novalis), ISBN 0–8146–2480–4 (LP), ISBN 0–85574–063–9 (Dwyer) MT88 .R45 1996

Practical suggestions regarding the use of music within the worship service of Catholic churches. Covers selection of music, sources for liturgical music, physical placement of musicians, and introduction of new music.

Illustrations; glossary of terms; briefly annotated bibliography of thirteen writings.

557. Sotak, Diana Kodner. *Handbook for Cantors.* Chicago.: Liturgy Training Publications, 1988. vii, 85 p. ISBN 0–930467–89–2 M2117.S83 H3

Manual for cantors in Catholic churches. Covers psalms and other cantor repertory, ways of singing the psalms with and without a cantor, teaching music to the assembly, physical communication and animation of the assembly, the Eucharistic liturgy, singing technique, interpretation, and other matters. Numerous musical examples; exercises at the end of sections; recommended readings at the end of chapters.

See also: Funk, Virgil C., ed. *Initiation and Its Seasons* (item 275); Funk, Virgil C., ed. *The Pastoral Musician* (item 276); Funk, Virgil C., ed. *The Singing Assembly* (item 277)

Church of Christ

558. Jackson, James L. *Church Music Handbook: A Cappella Singing.* Nashville, Tenn.: James L. Jackson, 1983. 100 p. BV290 .J33 1983

Covers music fundamentals (pitch, clef, scales, etc.), suggestions for leading a choir, and selecting music and planning worship assemblies. Music examples, illustrations, and tables.

Episcopal/Anglican

559. *The Episcopal Choirmaster's Handbook.* Milwaukee, Wis.: The Living Church Foundation, 1956–. (Annual). ML3166 E6

Handbook provides service outline for Rite I and Rite II along with prayers, readings, and selections of music. Prefatory pages provide valuable information, including a table of canticles, titles and terminology, notes on the canticles for the Daily Offices, information on A Festival of Lessons and Carols, and an annotated bibliography of more than fifty useful writings and sources for music. Index of first lines of hymns.

560. Episcopal Church. Diocese of Connecticut. Music Commission. *When a Church Calls a Musician: A Handbook for Parish Churches and Pastoral Musicians.* 3rd ed. Hartford, Conn.: Music Commission, the Episcopal Diocese of Connecticut, 1994. ii, 28 p. ISBN 0–9636919–1–0 ML3795 .E65 1994

Handbook primarily devoted to personnel concerns (hiring practices, preparing a position description, determining salary, etc.). The fourteenth section entitled “Resources” lists tools for service planning. The appendix provides a listing of national organizations and their publications, music distributors, music publishers, and lending libraries, among other helpful tools.

561. Hatchett, Marion J. *A Guide to the Practice of Church Music.* New York: Church Hymnal Corporation, 1989. 250 p. ML3166 .H16 1989; reprint, New York: Church Publishing, 1998. ML3166 .H16 1998

Extensive revision and expansion of author’s *The Book of Occasional Services: Conforming to the General Convention, 1988* (1988). Geared toward the Episcopal Church. Covers musical ministries, types of music (hymns, psalms, service music, anthems, and instrumental voluntaries), educating and inspiring the congregation, and planning music for the rites of the *Book of Common Prayer* and the *Book of Occasional Services.* Three appendixes: (1) list of approximately three hundred descants, fauxbourdons, varied harmonizations, varied accompaniments, suggested alternative treatments of hymns, and hymns scored for other instruments found in *The Hymnal 1982*; (2) metrical index of tunes in *The Hymnal 1982* with the first lines of the texts; and (3) checklists for planning no less than fifty-one services. Musical examples.

562. Keiser, Marilyn J. *Teaching Music in the Small Church.* New York: Church Hymnal Corporation, 1983. 64 p. ISBN 0–89869–102–8 MT88.K45 T4

Based on the author’s experience as Music Consultant for the Diocese of Western North Carolina. Provides instruction for teaching new music and for improving choir and congregational singing within small churches. Numerous musical examples; documented with endnotes; bibliography of forty-nine recommended anthems for small choirs.

Lutheran

563. Association of Lutheran Church Musicians. *Guide to Basic Resources for the Lutheran Church Musician: A Project of the Educational Concerns Committee.* Preface by Carl Schalk, ed. St. Louis, Mo.: MorningStar Music Publishers for the Association of Lutheran Church Musicians, 1995. 61 p. ISBN 0–944529–13–5 ML3168 .A77 1995

Handbook intended for musicians that have “little or no background or understanding of liturgical worship in the Lutheran tradition.” Presents brief historical information about Lutheran worship. Addresses selecting an organ, the use of instruments in worship, singing the propers, and other related topics. The section titled “Service Playing: Techniques and Literature” offers an annotated, classified bibliography of six writings, a few videos and cassettes, and approximately fifty collections of music covering basic organ literature for worship, easier liturgy and hymn accompaniments, and music for organ with instruments. Section titled "The Choir and the Choir Director" lists thirty-one annotated writings and a few videos on conducting, voice building, early childhood music education, children's choirs, youth choirs, and adult choirs.

564. Caemmerer, Richard R. "The Congregational Hymn as the Living Voice of the Gospel" in *The Musical Heritage of the Lutheran Church,* vol. V, ed. Theodore Hoelty-Nickel. St. Louis, Mo.: Concordia, 1959. pp. 166–177. ML3168 .M8 1953–1957

Essay presented at the Valparaiso University Church Music Seminary during its annual sessions from 1953 to 1957. Thesis: "a hymn in a service of Christian worship is a means by which each worshiper speaks the living Word of God to each other one." Examines biblical references to hymns, summarizes the essential principles of the hymn, and applies these principles to the spiritual growth of the congregation.

565. Halter, Carl, and Carl Schalk, eds. *A Handbook of Church Music.* St. Louis, Mo.: Concordia, 1978. 303 p. ISBN 0–570–01316-X ML3168 .H33

Companion to *Key Words in Church Music: Definition Essays on Concepts, Practices, and Movements of Thought in Church Music* (item 184), also edited by Schalk, though the *Handbook* concerns only Lutheran worship music. Each book is cross referenced to the other. Introduction, "Music in Lutheran Worship: An Affirmation," by Halter and Schalk, followed by seven chapters by various authors: "The Liturgical Life of the Church" (E. L. Brand); "Sketches of Lutheran Worship" (Schalk); "The Music of the Congregation" (L. G. Nuechterlein); "The Music of the Choir" (C. R. Messerli); "The Music of Instruments" (H. Gotsch and E. W. Klammer); "The Pastor and the Church Musician" (A. R. Kretzmann); and "Music in the Church Today: An Appraisal" (R. Hillert). Musical examples, illustrations, and photographs; extensive annotated bibliography prepared by Messerli of approximately two hundred items; tables of the Church Year, the Mass, and various Lutheran Orders; no index.

566. Sedio, Mark, Gerald Coleman, Sharon Sasse Silleck, Vern Gundermann, David Christian, Thomas Zehnder, Paul Bouman, and Elizabeth Gotsch. *Living Voice of the Gospel: Dimensions in Wholeness for the Church Musician.* Saint Louis, Mo.: Concordia, 1996. 80 p. ISBN 0–570–01353–4 MT88 .L58 1996

Essays by Lutheran authors. Addresses creative/artistic, pastoral, historical, ecclesiastical, and theological dimensions of the music ministry.

Mennonite

567. Neufeld, Bernie, ed. *Music in Worship: A Mennonite Perspective.* Scottdale, Pa.: Herald Press; Newton, Kans.: Faith & Life Press, 1998. 260 p. ISBN 0–8361–9459–4 ML3169 .M87 1998

Neufeld, Bernie, ed. *Music in Worship: A Mennonite Perspective.* Boulder, Colorado, 2001. 260 p. ISBN 0–585–37272–1 Computer file

Fourteen essays on the role of music in worship from a Mennonite perspective. Essays: "Worship: True to Jesus" (E. Kreider); "An Anabaptist Perspective on Music in Worship" (J. Rempel); "Crossing the Border: Music as Traveler" (Neufeld); "Global Music for the Churches" (M. K. Oyer); "Contemporary Church Music Issues" (C. Longhurst); "Congregational Singing as a Pastor Sees It" (G. Harder); "Anticipating God-Presence: Recovering a Primary Essential for Worship" (G. D. Wiebe); "Creative Hymn Singing" (M. H. Hamm); "Children and Music Ministry" (J-E. Grunau); "And What Shall We Do with the Choir?" (K. Nafziger); "Church Music as Icon: Vehicle of Worship or Idol" (D. Bartel); "Silencing the Voice of the People: Effects of Changing Sanctuary Designs" (J. E. Kreider); "The Composer as Preacher" (L. J. Enns); and "The Hymn Text Writer Facing the Twenty-First Century" (J. Janzen). Also published as an e-book. Mode of access: World Wide Web.

568. Schmidt, Orlando. *Church Music and Worship among the Mennonites.* Newton, Kan.: Faith and Life Press; Scottdale, Pa.: Mennonite Publishing House, 1981. 40 p. ISBN 0–87303–047–8 ML3169 .S35

Booklet prepared for those involved in the planning and leading of the Mennonite worship service. Three brief chapters discuss the formalization of Mennonite church and worship music, provide guidelines for church and worship music, and address worship in the late twentieth century. Documented with endnotes.

United Methodist

569. Bone, David L., and Mary J. Scifres. *The United Methodist Music and Worship Planner, 1994–1995.* Nashville, Tenn.: Abingdon Press, 1994. 137 p. ISBN 0–687–43175–1 BX8337 .B66 1994

Previously published in 1992 and 1993 as *The United Methodist Music and Worship Planner, 1992–1993* and *The United Methodist Music and Worship Planner, 1993–1994*, respectively. A worship planning guide for pastors, musicians, and worship leaders in the United Methodist Church.

Suggestions for hymns, psalters, keyboard music, vocal solos, anthems, and hymn anthems for use through the church year.

Quakers

570. Riney, Cecil J. "The Emergence and Development of a Ministry of Music within the Society of Friends." D.M.A. dissertation. Los Angeles: University of Southern California, 1963. iv, 346 p.

Discusses the history of the Society of Friends (Quakers) and the development of music within the movement and examines current musical practices as of 1963. Tables; bibliography of more than one hundred writings; several appendices, including a bibliography of approximately two hundred anthems arranged according to level of difficulty suitable for church choirs and a bibliography of approximately 120 anthems arranged according to performing forces (unison, two-part, and combined choirs) suitable for youth choirs.

571. Vaughn, Carol Cline. "The Role of Music within the Friends Church: 1962–1982." Ed.D. dissertation. Tex.: University of Houston, 1990. xii, 226 p.

Examines the role of music in the worship services, educational institutions, and mission fields of the Society of Friends (Quakers) in the United States from 1962 to 1982. A continuation of Cecil J. Riney's study, "The Emergence and Development of a Ministry of Music within the Society of Friends" (item 570), which discusses music within Quakerism from the beginning of the Friends Movement in 1647. Tables, illustrations, and photographs; bibliography of nearly seventy writings and interviews and eight music scores and collections of music.

Reorganized Church of Jesus Christ of Latter-Day Saints

572. Revell, Roger A. *Hymns in Worship: A Guide to Hymns of the Saints.* Foreword by Wallace B. Smith, Duane E. Couey, and Howard S. Sheehy. Independence, Mo.: Herald Publishing House, 1981. 224 p. ISBN 0–8309–0335–6 ML3174 .R48

Serves as a guide to *Hymns of the Saints*, hymnal of the Reorganized Church of Jesus Christ of Latter-Day Saints. Provides sixty-four-page overview of hymn form and 108-page discussion of the use of hymns in worship (i.e., selecting hymns, learning new hymns, noncongregational uses of hymns, etc.). Glossary of terms; musical examples, photos, and tables; expansive index.

CHOIRS AND CHORAL MUSIC

573. Anderson, Jean. *You Can Have More than Music: Practical Helps for Planning, Directing, Leading, and Strengthening the Adult Choir.* Nashville, Tenn.: Abingdon Press, 1990. 31 p. ISBN 0–687–04607–6 MT88 .A57 1990

 A handbook for organizing and directing an adult church choir. Illustrations and musical examples.

574. Barfoot, Phil. *The Ultimate Idea Book for Music Ministry!: 500 Great Ideas from 107 Outstanding Music Ministries.* S.l.: Word Music, 1999. 464 p. ML3160 .B37 1999

 Handbook that offers five-hundred-plus ideas for the music minister in the areas of rehearsal, recruitment, promotion, worship service, seasonal/non-seasonal productions, youth choir, senior adult choir, staff relationships, balancing family and career, etc. Draws from the experience of 107 Protestant music ministers (profiles included). Provides a recommended listing of twenty-five octavos, twenty-five choral collections, and twenty-five musicals. Provides resources relating to banners, drama, publications, Christmas music, celebrations, and youth choir. Illustrations.

575. Cabaniss, Mark, and Janet Klevberg, compilers. *The ChoirBuilder Book: A Manual for Building Successful Music Ministries.* Milwaukee, Wis.: Brookfield Press, 1997. 39 p. MT88 .C53 1997

 Eleven brief chapters by various authors concerning church choirs: "Making Super Choirs with Everyday Singers" (T. Cleveland); "Blended Worship" (M. Acker); "Drama in the Church: To Do or Not to Do?" (D. Craig-Claar); "Growing Your Children's Choir" (P. Andrews); "Ten Great Ideas for Recruitment" (F. Robertson); "MIDI for the Church Musician" (R. Nagy); "Building a Successful Instrumental Program" (J. G. Gage); "The Efficient and Effective Choir Rehearsal" (C. H. Causey); "Choosing Quality Repertoire" (J. Purifoy); "Healthy Staff Relationships" (Cabaniss); and "Avoiding Burnout" (W. H. Rayborn).

576. Cronin, Deborah K. *O for a Dozen Tongues to Sing: Music Ministry with Small Choirs.* Nashville, Tenn.: Abingdon Press, 1996. 80 p. ISBN 0–687–01005–5 ML3001 .C84 1996

 Concerns musical realities facing churches with small memberships. Provides practical advice for recruiting singers, choosing music, conducting rehearsals, and exploring creative options such as the use of synthesizers, organizing singers into duets, trios, and quartets, and alternative worship services. Musical examples; no bibliography; no index.

577. Donathan, David F. *How Does Your Choir Grow?* Nashville, Tenn.: Abingdon Press, 1995. 63 p. ISBN 0–687–01075–6 MT88 .D66 1995

Addresses choir organization and development within the Protestant Church. Illustrations.

578. Edwards, Randy. *Revealing Riches & Building Lives: Youth Choir Ministry in the New Millennium.* St. Louis, Mo.: MorningStar Music Publishers, 2000. viii, 417 p. ISBN 0–944529–31–3 MT88 .E3 2000

Resource guide for conductors of Protestant church youth choirs. Practical advice on organizing and building a youth choir. Latter third of the book offers one hundred "effective ideas," an annotated bibliography of more than eighty Web sites, one hundred recommended readings for a youth choir director, twenty recommended movies "to gain more insight into the current youth culture," and nearly 340 choral works, providing difficulty ratings, voicing/scoring, publication information, and helpful notes.

579. Fortunato, Connie. *Children's Music Ministry: A Guide to Philosophy and Practice.* Foreword by Ray Robinson. Elgin, Ill.: David C. Cook, 1981. xvi, 217 p. ISBN 0–89191–341–6 MT88 .F65

Organized into three main parts: biblical perspectives, educational perspectives, and contemporary challenges. Part I, biblical perspectives, covers musical function in the Bible (music as worship, as education, and as evangelism), various musical types in the Bible (psalm singing, hymn singing, and spiritual songs), musical structure in the Bible (leadership, specialized participation, and congregational participation), and development of biblical music (development of psalmody, hymnody, spiritual songs, and biblical structure). Part II, educational perspectives, discusses cognitive, affective, and psychomotor development of children, setting instructional objectives, and motivating children, and offers practical advice regarding teaching methodology. Part III, contemporary challenges, deals with organizing children's music program, music in schools, and conducting successful rehearsals and performances. Musical examples, illustrations, and photos; bibliographies at the end of each section for a total of about forty-five writings; expansive index.

580. Funk, Virgil C., ed. *Children, Liturgy, and Music.* Washington, D.C.: Pastoral Press, 1990. vi, 130 p. ISBN 0–912405–73–2 ML3002 .P25 1981 v.2

Fifteen chapters written by various authors. Chapters dedicated to music are: "Music, Gestures, and Pictures—All for Children" (N. Chvatal); "Celebrating the Sophisticated Song of Children" (L. Gwozdz); "Enthusiasm in Children's Choirs" (D. G. Campbell); "What about Children's Choirs?" (D. Hruby); "Music Education and the Parish: A Dream" (R. Haas); "Music and Education: The Sky's the Limit" (L. Bufano); and "Should a Musician Offer a Religious Educator a Hymnbook?" (E. Downing).

581. Hansen, Linda B. *And the Song Goes On . . . : Older Adults in Music Ministry.* With Betsy Pittard Styles, consultant. Nashville, Tenn.: Abingdon Press, 1995. 79 p. ISBN 0–687–01147–7 MT88 .H295 1995

Manual intended for the inexperienced choir director of a senior adult church choir. Covers music and some liturgy. Musical examples; bibliography of eleven writings.

582. Jacobs, Ruth Krehbiel. *The Successful Children's Choir.* S.1.: H. T. FitzSimons, 1995. 63 p. ISBN 0–9646552–2–5 MT915.J24 S9 1995

Brief guide to organizing and training a children's choir, with special emphasis on religious education. Classified list of approximately two hundred sacred works for children's choirs and forty-three recommended writings. No index.

583. Link, John V., and Gerald Ware. *Keys to a Successful Youth Choir Ministry.* Nashville, Tenn.: Church Street Press, 1997. 94 p. ISBN 0–7673–3456–6 BV4427 .L55 1997

Directed toward the choir director of a church youth choir. Focuses on music and ministry, covering the selection of music, singing exercises, and rehearsals, as well as self motivation, encouragement of youth choir members, and outreach programs. Illustrations.

584. McDonald, Thomas John. "The Reason We Sing: A Choral Based Strategy for Music Ministry in the 90's." Ph.D. dissertation. Montpelier, Vt.: Union Institute, 1992. v, 154 p.

Unfolds a philosophy of music ministry in relationship to choral music. Illustrations and tables; documented with endnotes; glossary of terms; classified bibliography of about one hundred recommended anthems for adult, youth, and children choirs.

585. McRae, Shirley W. *Directing the Children's Choir: A Comprehensive Resource.* New York: Schirmer Books, 1991. xiv, 232 p. ISBN 0–02–871785–6 MT915 .M49 1991

Eight chapters with cross references. Provides practical advice for promoting and organizing a children's choir; presents the historical, theological, and educational basis for Christian children's choirs; examines characteristics of children, ages 4–11, providing insights and suggesting goals; presents the Kodály and Orff approaches to singing and methods of applying each to children's choirs; discusses organizational, musical, and budgetary considerations; gives pedagogical advice for working with the child's voice; offers guidelines for rehearsal; and makes suggestions for enriching the program. Five appendixes: list of professional organizations; bibliography of over one hundred books, collections of music, journals, and videocassettes; additional bibliography of approximately twenty recordings of children's choirs, audiovisuals, books for children with addresses of musical instrument vendors; a guide to abbreviations in Orff arrangements; and an essay titled "The Church Musician and the Copyright Law." Photos, musical examples, graphs, illustrations; expansive index.

586. Miller, Andrea Wells, ed. *A Choir Director's Handbook.* Waco, Tex.: Word, 1981. 211 p. ISBN 0–8499–8160–3 MT88 .C45x

Nineteen chapters on choral music ministry. Chapters: "How to Achieve Excitement and Momentum in a Choir Program and Keep It" (J. R. Wyatt); "Beyond the Choir Loft and Back Again: Evangelism through Music" (S. Salsbury); "Special Projects—Where Does the Money Come From?" (E. M. Craig); "Good News for Volunteer Choir Directors!" (M. M. Moore); "Developing Good Repertoire and a Balanced Choral Library" (J. Purifoy); "Planning the Worship Service" (R. D. Dinwiddie); "An Orchestra in Your Church—You Can Do It!" (A. Edwards); "The Impact of Small Ensembles in Your Music Program" (D. S. Blackburn); "Revitalizing Your Choir Rehearsals" (K. Kaiser); "The Choir Director as Producer" (W. K. Andress); "Taped Accompaniment—Friend or Foe?" (D. Blakeney); "Making a Musical Happen with Drama" (N. Knighton); "'Show and Sing' with Slides" (L. White); "Going on the Road" (C. Floria); "Publicity Basics—Covering the Bases" (Miller); "What Pastors Wish Choir Directors Knew" (L. Ogilvie); "What Accompanists Wish Choir Directors Knew" (O. Griffin); "What Teenagers Wish Choir Directors Knew" (P. Terry); and "What Choir Members Wish Choir Directors Knew" (K. Stokes). Musical examples and illustrations included in some chapters; general index.

587. Nordin, Dayton W. *How to Organize and Direct the Church Choir.* West Nyack, N.Y.: Parker, 1973. 214 p. ISBN 0–13–425207–1 MT88 .N74

Ten-chapter discussion of planning, organizing, managing, and directing a successful church music program. Includes an annotated bibliography, "Church Music for Every Occasion," with approximately 150 choral works classified by church year. Entries provide title, composer, publisher, and required voices. Expansive index.

588. Page, Sue Ellen. *Hearts and Hands and Voices: Growing in Faith through Choral Music.* Foreword by Helen Kemp. Tarzana, Calif.: H. T. FitzSimons, 1995. xiv, 196 p. ISBN 0–9646552–0–9, ISBN 0–9646552–1–7 (pbk.) MT915 .P34 1995

Based on principles presented in Ruth Krehbiel Jacobs's *The Successful Children's Choir* (3rd. ed., Choir Publications; H. T. FitzSimons, 1948). Explores recent trends in church music. Primarily devoted to discussions relating to conducting, rehearsals, and repertoire of the children's church choir. Musical examples, illustrations, and photographs; various appendixes provide annotated bibliographies of recommended choral works (seventy for elementary and upper elementary choirs; twenty-six extended sacred choral works; eighty-seven hymns for young children); classified bibliography of approximately 120 writings, forty periodical titles and organizations, and twenty video recordings; expansive index.

589. Pfautsch, Lloyd. *Choral Therapy: Techniques and Exercises for the Church Choir.* Nashville, Tenn.: Abingdon Press, 1994. 88 p. ISBN 0–687–06510–0 MT875 .P44 1994

Diction and rehearsal techniques for the church choir. Much of the content would apply to choral music in general. Glossary of terms; musical examples and illustrations.

590. Schultz, Ralph C. *Leading the Choir: A Guide for the Inexperienced Choir Director.* St. Louis, Mo.: Concordia, 1989, 1990. 32 p. + 1 videocassette. MT85 .S39 1990

Instruction book accompanied by a videocassette for the beginning church choir director. Covers placement of voices, rehearsals, vocal development, importance of diction, and conducting patterns. Musical examples, illustrations, and photos; bibliography of four writings.

See also: Bobbitt, Paul, and Gerald Armstrong. *The Care and Feeding of Youth Choirs* (item 534); Brownstead, Frank, and Pat McCollam. *The Volunteer Choir* (item 543); Easterling, R. B. *Church Music for Youth* (item 535); Vaught, W. Lyndel. *Senior Adult Choir Ministry* (item 541)

CONTEMPORARY WORSHIP MUSIC

591. Arvin, Reed, ed. *The Inside Track to—Getting Started in Christian Music.* Eugene, Ore.: Harvest House, 2000. 350 p. ISBN 0–7369–0267–8 MT67 .G37 2000

Fourteen essays by various authors on life as a Christian music artist and/or songwriter in the commercial music field. Originally published as *The AGMA Music Curriculum* (Academy of Gospel Music Arts, 1999). Essays: "Music, Ministry, and Fame" (Arvin); "The Role of Music in Worship" (H. Best); "The Creative Christian Life, Part 1" (C. Peacock); "The Creative Christian Life, Part 2" (Arvin); "The Christian in Secular Music"

(M. Roe); "A History of Contemporary Christian Music" (P. Kavanaugh); "The Artist/A&R Director Relationship" (D. Posthuma); "The Artist/Church Relationship" (S. Smith, S. Green); "Booking and Self-Management" (B. Connolly); "Recording the Independent Project" (Arvin); "Initial Steps to Building a Successful Song" (M. Becker); "Poetic Devices" (R. Sterling); "Writing is Rewriting" (J. Lindsey); and "The Songwriter/Publisher Relationship" (D. Cason). Documented with endnotes; photos of contributors.

592. Barrett, Bob. *Contemporary Music Styles: The Worship Band's Guide to Excellence.* Mission Viejo, Calif.: Taylor Made Music, 1996. 234 p. + 1 compact disc. MT170 .B35 1996

Manual covering the fundamentals of contemporary music styles, including the rhythm section, contemporary pop/rock, ballads, country, gospel, Latin, funk/R&B, and playing hymns in a contemporary service. Numerous musical examples. Accompanied by a compact disc recording of music exhibiting many of the styles presented in the text.

593. Flather, Douglas R., and Tami Flather. *The Praise and Worship Team Instant Tune-Up!* Grand Rapids, Mich.: Zondervan, 2002. 143 p. ISBN 0–310–24232–0 MT88 .F53 2002

Offers practical tips for organization, rehearsal, and performance of a praise and worship group. Covers contemporary harmonies, use of dynamics, mixed instruments, singing, among other topics. Numerous illustrations and tables.

594. Hurst, Lynn. *Changing Your Tune!: The Musician's Handbook for Creating Contemporary Worship.* Nashville, Tenn.: Abingdon Press, 1999. 143 p. ISBN 0–687–02297–5 MT88 .H877 1999

Considers music leadership in Protestant churches. Covers worship space, sound and projection systems, necessary licenses, organizing and rehearsing worship and praise teams, designing the service, rehearsal, advertising, budget, and evaluating the service. Few musical examples; glossary of terms relating to sound and projection systems.

595. Hurst, Lynn, and Sherrell Boles. *Praise Now!: Ready-to-Use Services for Contemporary Worship.* Nashville, Tenn.: Abingdon Press, 2000. 88 p. ISBN 0–687–09080–6 BV198 .H87 2000

Twenty thematic services for public worship. Provides a variety of materials, including scripts with order and approximate times for each component of worship, list of music resources for each service, skits and liturgies written for each specific service; other optional resources for music, recordings, dramas, and movies, and a sermon guide. Photos; index of song titles.

596. Liesch, Barry Wayne. *The New Worship: Straight Talk on Music and the Church.* Expanded ed. Foreword by Donald P. Hustad. Grand Rapids, Mich.: Baker Books, 2001. 269 p. ISBN 0–8010–6356–6 BV290 .L54 2001

Guide to contemporary worship practices in Protestant churches. Accompanying software available on modulation and improvisation; supplemental materials available via the World Wide Web. Illustrations and tables; documented with endnotes; expansive index.

597. Schneider, Kent E. *The Creative Musician in the Church.* West Lafayette, Ind.: Center for Contemporary Celebration, 1976. iv, 203 p. ML3001.S36x

Primarily for performers of contemporary church music. Reviews contemporary practices over a twenty-year period, addresses relationship of music to the liturgy, offers insight into composing lyrics and music, and provides practical advice for the church musician. Musical examples and photographs; bibliographies of recommended readings follow some chapters; index of names and expansive index of subjects.

HYMNS AND HYMNODY

598. Dakers, Lionel. *Choosing and Using Hymns.* Foreword by Robert Cantuar. London: Mowbray, 1985. xi, 96 p. ISBN 0–264–67034–5 BV370 .D24 1985

Handbook that offers practical advice on choice and performance of hymns. Dakers, a British scholar, insists that the content of this book "has no denominational or geographical boundaries or slant, but is directed towards *all* concerned in worship." Musical examples; bibliography of fifteen writings.

599. Fisher, Tim. *Harmony at Home: Straight Answers to Help You Build Healthy Music Standards.* Greenville, S.C.: Sacred Music Services, 1999. v, 203 p.

In two parts. Part I, comprising the first quarter of the book, is described by the author as "a challenge to each family to seek the blessing of music in the home through a discussion of why we sing, how we use hymns and hymnals, how we teach music standards to our children, and tools to aid us in this ministry." Provides a partially-annotated bibliography of writings, scores, recordings, and other miscellaneous materials. Part II, comprising the remainder of the book, consists of twenty commonly-asked questions with detailed answers. Few musical examples; scripture index.

INSTRUMENTAL MUSIC

General Works

600. Adams, Jere V., and Gerald P. Armstrong, eds. *The Church Instrumental Ministry: A Practical Guide.* Nashville, Tenn.: Convention Press, 1989. 31 p. ML3869.C5 A4 1989

Thirteen two-page chapters by various authors. Chapters: "Why Bother with Instruments?" (D. Danner); "Beginning a Graded Instrumental Program" (L. Poquette); "How to Find Instrumentalists" (C. Krause); "The Why's and How's of Warming Up" (B. Maples); "Directing a Junior High Instrumental Ensemble" (B. Johnson); "Working with Senior High Instrumentalists" (J. Hanbery); "The 'Has Been' Adult Instrumental Ensemble" (D. Allen); "The Choir of Brass" (D. Smith); "Conducting a Woodwind Rehearsal" (R. Ford); "Conducting Rehearsals with the Mixed Instrumental Ensemble" (J. King); "Planning an Instrumental Concert" (Armstrong); "Instruments and Worship Planning" (B. Morris); and "Using Professional Musicians" (P. Tobias). Photos; learning activities provided for each chapter.

601. Haugen, Marty. *Instrumentation and the Liturgical Ensemble.* Introduction by Rembert G. Weakland. Chicago.: G. I. A. Publications, 1991. xi, 208 p. + 2 audio cassettes. ML3001 .H38 1991

In the author's words, "This book addresses how musical instruments are used in various combinations to support and nurture the assembly's prayer." Eleven chapters cover: the task of the parish musician, the value of instrumentation in worship, the liturgical ensemble in worship, the pipe organ as a model for worship instrumentation, the pastoral dimension of instruments, the historical use of instrumentation in liturgy, public address systems and acoustical environment, instrumentation in the real parish world, and separate chapters on guitar, piano, organ, electronic keyboards, bass instruments, woodwinds/strings/brass, and percussion. Illustrations and numerous musical examples; partially annotated bibliography of nearly twenty writings; expansive index.

Handbell Music

602. Folkening, John. *Handbells in the Liturgical Service.* Saint Louis, Mo.: Concordia, 1984. 45 p. ISBN 0–570–01328–3 MT711 .F64

Practical handbook concerning the use of handbells in the worship service. Provides historical and technical information. Several appendixes specifically address the use of handbells in Lutheran worship. Musical examples, illustrations, and photographs; bibliography of four writings.

603. Frazier, James, Kermit Junkert, Donald Shier, Robert Strusinski, and Bonita Wurscher. *Handbells in the Liturgy: A Practical Guide for the Use of Handbells in Liturgical Worship.* Saint Louis, Mo.: Concordia, 1994. 120 p. MT710 .H3 1994

    Manual to "assist in integrating handbells into more liturgical forms of worship." Geared toward Catholic, Lutheran, and Episcopal churches. Covers organizing a handbell program, handbell techniques, writing handbell accompaniments, and selecting seasonal music. Musical examples and photos; documented with endnotes; bibliography of sixteen recommended readings.

Keyboard Music

604. Cherwien, David. *Let the People Sing! A Keyboardist's Creative and Practical Guide to Engaging God's People in Meaningful Song.* St. Louis, Mo.: Concordia, 1997. 179 p. ISBN 0–570–01354–2 MT190 .C4 1997

    Guide for the church keyboardist. In two parts: (1) tips for song leading, hymn playing, and accompaniment for a variety of styles of music, and (2) improvisation at the keyboard. Numerous musical examples; indexes for musical examples.

605. Lovelace, Austin C. *The Organist and Hymn Playing.* Rev. ed. Carol Stream, Ill.: Agape, 1981. iv, 61 p. ISBN 0–916642–16-X MT180 .L69 1981

    Originally published in 1962, revised in 1981. An instruction book for hymn playing on organ. Covers pedaling, articulation and touch, introducing the hymn, tempos, "amens," hymn forms, registration, variety in hymn playing and singing, and free harmonizations. Numerous musical examples; bibliography of approximately seventy free accompaniments of hymns. Bibliography of twenty-six writings.

606. Riddle, Pauline. "The Development of a Basic Foundation in Church Organ Technique for the Beginning Organist." D.Mus.Ed. dissertation. Norman: University of Oklahoma, 1972. v, 260 p.

    Exercises composed by the author to teach attack and release of a note, repeated notes, glissando, and substitution. Graphs and musical examples; bibliography of approximately forty writings and fifteen hymnals.

607. Riddle, Pauline. *Five Practical Lessons for Church Organists.* Nashville, Tenn.: Convention Press, 1985. 31 p. MT185 .R52 1985

    Lessons: fundamentals of the organ; manual techniques for hymn playing; pedal technique; fundamentals of music theory; and cadences. Illustrations and numerous musical examples; index of hymn tunes in the text.

MIDI

608. Heinzman, David Lee. *MIDI Goes to Church: An Introduction and Practical Guide to Musical Instrument Digital Interface for the Church Musician.* Foreword by Dale Jergenson. Van Nuys, Calif.: Laurendale Associates, 1991. 84 p. MT723 .H45 1991

Four chapters: Chapter 1 handles basics about Musical Instrument Digital Interface (MIDI), such as defining MIDI, explaining MIDI channels and modes, what to look for in MIDI equipment, MIDI keyboards and sequencers, and MIDI systems on church organs. Chapter 2 discusses integrating MIDI into worship, including discussion about the MIDI organ and use of MIDI for accompaniment with the choir and in contemporary Christian music. Chapter 3 briefly describes MIDI sequencing. Chapter 4 handles MIDI technical matters, including MIDI plugs, ports, and cables, and various MIDI equipment configurations. Glossary of terms; musical examples and illustrations. Accompanied by a video titled *MIDI Goes to Church, The Video.*

Orchestra Music

609. Adams, Jere V., and Gerald P. Armstrong, eds. *Orchestral Concepts in Today's Church.* Nashville, Tenn.: Convention Press, 1991. 83 p. MT730 .O73 1991

Text for a portion of the Church Study Course of the Baptist Sunday School Board. Eleven chapters by various authors: "Organizing the Church Orchestra" (J. E. Helman); "The Church Orchestra Built on a Traditional Concept" (L. W. Mayo, J. G. Gage); "The Ministry of the Church Orchestra" (L. Poquette); "The Orchestra in the Small Church" (G. M. D. Frink); "The Orchestra in the Large Church" (M. D. Johnson); "The Churchestra," which refers to an instrumental ensemble that plays in the church (D. Smith); "The Piano-Plus Orchestra" (B. Walters); "The 7-Plus Orchestra," which refers to an ensemble that has a basic core of seven specific instruments, though substitute instruments may be used (C. Kirkland); "The Rhythm Section-Based Orchestra" (D. S. Winkler); "Adapting Wind Ensemble Music to the Church Orchestra" (C. Krause); and "Soloists and Ensembles from the Church Orchestra" (Armstrong). Musical examples, charts, and illustrations.

610. Frink, George M. D. *Today's Church Orchestra.* 2nd ed. Charleston, S.C.: Carol Press, 1997. 82 p. ML3001 .F74 1997

Purpose: "to trace something of the historical development of the church orchestra, give some guidelines concerning the starting of one in a local church, and give some insight into the problems which the arranger faces in preparing music materials for such an organization." Musical examples and illustrations; brief bibliography of writings, a handful of scores for choir and orchestra, and approximately forty scores for orchestra.

611. Frink, George M. D. *What Do You Know about a Church Orchestra: What's Different about It?* Charleston, S.C.: Carol Press, 1990. 19 p. MT70.F74 W4 1990

Essay-length. Provides helpful tips for organizing a church orchestra. Also, offers advice for arranging and orchestrating for the ensemble. Illustrations.

612. Johansson, Calvin M. *Discipling Music Ministry: Twenty-First Century Directions.* Peabody, Mass.: Hendrickson Publishers, 1992. vi, 169 p. ISBN 0–943575–52–4 ML3001 .J63 1992

Serves as a companion to Johansson's *Music & Ministry: A Biblical Counterpoint* (item 613). Addresses the role of church music within the discipling process. Musical examples; documented with endnotes.

613. Johansson, Calvin M. *Music & Ministry: A Biblical Counterpoint.* 2nd ed. Peabody, Mass.: Hendrickson Publishers, 1998. xi, 194 p. ISBN 1–56563–361-X (pbk.) ML3869 .J63 1998

Reviews "biblical principles foundational to music ministry." Bibliography of approximately 120 writings; index.

SONG LEADING AND CONGREGATIONAL SINGING

614. Boyd, Jack. *Leading the Lord's Singing.* Abilene, Tex.: Quality Publications, 1981. 204 p. ISBN 0–89137–603–8 ML3869 .B69

A book on song leading. Covers types of religious music (worship, instructional, and recreational music), basic definitions (psalms, hymns, spiritual songs, tune, harmonization, and text), skills of the song leader, choosing hymns, interpreting hymns, editing and modifying music, training the congregation, rehearsing, and directing. Musical examples, tables, illustrations, and photos; bibliography of more than fifty writings.

615. Boyd, Jack. *Leading the Lord's Worship.* Nashville, Tenn.: Praise Press/ Power Source Productions, 2001. vii, 308 p. BV290 .B69 2001

A handbook on song leading intended "to outline the more pressing problems in congregational singing." Reviews various types and styles of worship music, basic definitions, choosing hymns, interpreting music, editing and adapting hymns, and hand movements. Illustrations and music examples; classified, annotated bibliography of about seventy writings.

616. Campbell, James David. "Go into All the World and Sing the Gospel: A Strategy for Teaching Congregations through Music." D.Min. dissertation. Dayton, Ohio: United Theological Seminary, 1993. 246 p.

Explores the relationship between music and Christian education in worship. Includes an annotated bibliography of more than one hundred writings.

617. Farlee, Robert Buckley, and Eric P. Vollen, eds. *Leading the Church's Song*. Foreword by Paul Westermeyer. Minneapolis, Minn.: Augsburg Fortress, 1998. xii, 162 p. + 1 compact disc. ISBN 0–8066–3591–6 MT88 .L4 1998

Covers techniques for song leading (M. Mummert, M. Sedio, R. R. Webster, contributors), chant (R. Gallagher), and contemporary music (M. Glaeser. R. Webb). Separate chapter on North American church music (R. Knowles Wallace). Historical discussion is introductory at best. Numerous musical examples; bibliography of nearly ninety hymnals, hymnal supplements, and song collections, twelve hymnal companions, nearly forty writings on congregational song, thirteen writings on chant, ten writings on African-American music, and twenty-two writings on contemporary music; index of tune names and expansive general index.

618. Lehman, Glenn M. *You Can Lead Singing: A Song Leader's Manual.* Intercourse, Pa.: Good Books, 1995. 94 p. ISBN 1–56148–117–3 MT88 .L42 1995

Conducting manual for the Protestant church inexperienced song leader. Musical examples and illustrations; glossary of musical terms with index.

619. Parker, Alice. *Melodious Accord: Good Singing in Church.* Chicago, Ill.: Liturgy Training Publications, 1991. 122 p. ISBN 0–929650–43–3 ML3011 .P37 1991

Parker's ideas on the subject of good congregational singing. Written as a dialogue between the author and church musician. Musical examples and illustrations; documented with endnotes.

620. Sydnor, James Rawlings. *Hymns and Their Uses: A Guide to Improved Congregational Singing.* Carol Stream, Ill.: Agape, 1982. viii, 152 p. ISBN 0–916642–18–6 ML3000 .S94 1982

Divided into four sections: (1) the value and development of congregational singing; (2) the hymn and hymnal; (3) leadership of hymn singing; and (4) educating the congregation to sing hymns. Illustrations and musical examples; classified bibliography of forty-nine writings, nine periodical titles, and twelve hymnals; index.

621. Sydnor, James Rawlings. *Introducing a New Hymnal: How to Improve Congregational Singing.* Chicago.: G. I. A. Publications, 1989. ix, 132 p. ISBN 0–941050–19-X ML3111 .S952 1989

Two parts: (1) introducing a new hymnal and (2) how to improve congregational singing. Musical examples, illustrations, and tables; list of professional organizations; classified bibliography of nearly fifty writings; index.

622. Wren, Brian A. *Praying Twice: The Music and Words of Congregational Song.* Louisville, Ky.: Westminster John Knox Press, 2000. ix, 422 p. ISBN 0–664–25670–8 ML3270 .W74 2000

Electronic version: Boulder, Colorado: NetLibrary, 2000. URL: http://www.netLibrary.com/urlapi.asp?action=summary&v=1&bookid=41254

For pastors, worship leaders, and church musicians. Examines text and music of choruses, hymns, chants, and ritual songs. Bibliography of thirteen hymnals and nearly two hundred writings; separate indexes, some expansive, for scripture, names, subjects, and titles. Also published as an e-book, mode of access: World Wide Web.

See also: Reynolds, William Jensen. *Congregational Singing* (item 539)

IX

Tradition, Change, and Conflict

GENERAL WORKS

623. Ashton, Joseph N. *Music in Worship: The Use of Music in the Church Service.* Boston, Mass.: Pilgrim Press, 1943 (2nd ed., Pilgrim Press, 1943; 3rd ed., Pilgrim Press, 1943; 4th ed., Pilgrim Press, 1947). ML3001.A82 M8; 2nd ed. reprint, Bristol, Ind.: Wyndham Hall Press, 2001. New introduction and edited by Peter E. Roussakis. 232 p. ML3001.A82 M8 2001

     Dated, but philosophy presented is still embraced by more recent authors. Purpose: " . . . to set forth the essential function of church music and to present certain means of attaining it." Organized into two parts: (1) nine chapters address principles of church music, and (2) five chapters address application of the principles within music for the congregation, choir, and organ, and offers suggestions for the music director and organist. No bibliography; no index.

624. Ball, Louis, and Mary Charlotte Ball, eds. *On the State of Church Music, V: Church Music in the Twenty-First Century, A Symposium.* Jefferson City, Tenn.: Louis and Mary Charlotte Ball Institute of Church Music, and Center for Church Music, Carson-Newman College, 1997. 82 p. ML3106 .O5 1997

     Five essays. All contributors active in the United States. Essays: "A Sensuous God" (W. Hendricks); "Creator and Creativity: An Interchange" (Hendricks); "Theology, Seekers, Music, and Sensitivity" (H. M. Best); "When is Worship Worship?" (Best); and "The Development of Worship Styles" (R. Webber). Photos of contributors and some illustrations.

625. Ball, Louis, and Mary Charlotte Ball, eds. *On the State of Church Music, VI: Four Lectures on Church Music.* Jefferson City, Tenn.: Louis and Mary Charlotte Ball Institute of Church Music, and Center for Church Music, Carson-Newman College, 1998. 42 p. ML3106 .O5 1998

Four essays: "The Size of the Final Song" and "Church Music in the Service of Transcendence" (P. D. Duke); "We've a Story to Tell: Christian Music Publishing" (S. Lyon); and "The New Church Order" (D. Manley). Bibliographies follow some of the essays.

626. Ball, Louis, and Mary Charlotte Ball, eds. *On the State of Church Music, VIII: Three Lectures on Church Music Emphasizing Global Church Music.* Jefferson City, Tenn.: Louis and Mary Charlotte Ball Institute of Church Music, and Center for Church Music, Carson-Newman College, 2000. 68 p. ML3106 .O5 2000

Actually four essays by three authors emphasizing the use of global church music in church services in the United States Titles: "Issues in World Musics for the Church" (C. M. Hawn); "Thinking Globally, Singing Locally: How Can We Sing a Strange Song in the Lord's Land?: Moving from Dominance to Diversity" (T. W. Sharp); "Empiring or Empowering?: The Beginning of a New Music Ministry Era" (Sharp); and "A Short History of Church Instrumental Music—with a Southern Baptist Bias—in the Final Quarter of the 20th Century" (G. D. Smith). Bibliographies follow some of the essays.

627. Berglund, Robert D. *A Philosophy of Church Music.* Chicago.: Moody Press, 1985. x, 111 p. ISBN 0–8024–0279–8 ML3869 .B37 1985

Considers theological, philosophical, and psychological considerations in the development of a church music philosophy. Bibliography of eleven writings.

628. Chang, Jean Cho-hee. "Theological and Philosophical Perspectives on Today's Church Music Practice (Focusing on the Dichotomy of the Traditional/Contemporary)." D.C.M. dissertation. California: Claremont Graduate University, 1999. vi, 67 p.

Examines Protestant church music practice in relationship to the Old and New Testaments as well as other ancient and more contemporary Christian doctrines. Deals with contemporary Christian popular music and traditional church music. Musical examples; bibliography of more than fifty writings.

629. Collins, Mary, David Power, and Mellonee Burnim, eds. *Music and the Experience of God.* Edinburgh, Scotland: T. & T. Clark, 1989. xvi, 155 p. ISBN 0–567–30082-X ML197 .M817 1989

Collection of thirteen essays on church music. Part of *Concilium*, a multi-volume collection of writings on religious thought. Essays on church music in the United States: "Musical Traditions and Tensions in the American Synagogue" (L. A. Hoffman) "describes the background to the 'miscommunication' between rabbi and cantor based on a deep-seated ambivalence to music's expressive power [and] puts forward an alternative and non-judgmental model that allows for conflict but is rooted in the social contrast inherent to Jewish ecclesiology;" "Black Spirituals: A Theological Interpretation" (J. Cone) presents theological interpretations of selected African-American spiritual texts; "The Performance of Black Gospel Music as Transformation" (M. Burnim) "seeks to codify patterns of behavior which point to the existence of an underlying system of cultural values among Black Americans in the United States;" "Both In and Between: Women's Musical Roles in Ritual Life" (E. Koskoff) examines the musical role of women within the Jewish faith, Iroquois rites, and among other world cultures; "Word and Music in the Liturgy" (A. Nocent) and "The Path of Music" (J. Gelineau) address the relationship of music and text within the Catholic Church.

630. *Crisis in Church Music?: Proceedings of a Meeting on Church Music Conducted by The Liturgical Conference and The Church Music Association of America.* Washington, D.C.: The Liturgical Conference, 1967. 128 p. ML3000.1.L58 H4 1967

An expansion of *Harmony and Discord: An Open Forum on Church Music* (The Liturgical Conference, 1966). Twelve essays on church music: "Music and Liturgy in Evolution" (R. Weakland); "Sacred Music in the Teaching of the Church" (F. R. McManus); "The Theology of Liturgy According to Vatican II" (G. Diekmann); "Music in Lutheran Worship" (C. F. Schalk); "Leaning Right?" (F. P. Schmitt); "Church Music Today—The Center Position" (R. I. Blanchard); "In Praise of Joy—The Left Position" (C. A. Peloquin); "A View from the Far Left" (D. Fitzpatrick); "Music in the Church: A Declaration of Dependence" (B. Ulanov); "Facing Reality in the Liturgical Music Apostolate" (M. Theophane); "Music for the Vernacular Liturgy: Some Observations on Principles of Repertoire" (E. Lindusky); and "The Leader of Liturgical Song" (T. J. Reardon). Some essays documented with footnotes.

631. Davison, Archibald T. *Church Music: Illusion and Reality.* Cambridge, Mass.: University of Harvard Press, 1952 (rep. 1960, 1966). ix, 148 p. ML3000 .D3

Not a history of church music. Davison presents his philosophy of church music. Begins with a description of the nature of music and the nature of church music. Technical differences between sacred and secular music are outlined. Davison then addresses the dismal state of church music, with the concluding remark, "I see no prospect of extensive improvement in church music either Roman Catholic or Protestant until by purging it of its worldly substance we make of it something that is uniquely the music of worship." Annotated bibliography of seventy anthems, all by European composers; expansive index.

632. Dinwiddie, Richard. "Understanding God's Philosophy of Music." *Moody Monthly,* 74/3 (Nov. 1973): 48–54

The author writes, "God Himself has given us a rather comprehensive philosophy of music. It is flexible enough to allow for individual tastes and changing historical patterns of music. It is general enough so that one need not rethink his position every few years as a new style or approach is introduced." The author outlines this philosophy in six statements: (1) "We must have a balance in motivation," meaning to glorify God in word and deed; (2) "We must have a balance of direction," meaning to direct music to God, one another, and ourselves; (3) "Our music should have a balance of style," referring to the singing of different types of music, namely psalms, hymns, and spiritual songs; (4) "There is also a balance of function," meaning to encourage one another, sing of thanksgiving, to teach, to praise God, etc.; (5) "We must be particularly careful to preserve a balance between emotion and intellect," meaning to sing with the Spirit and the mind; and (6) "There must be a balance in performance," meaning to understand the abilities of the performing forces and its impact of the music on the congregation.

633. Doran, Carol, and Thomas H. Troeger. *Trouble at the Table: Gathering the Tribes for Worship.* Nashville, Tenn.: Abingdon Press, 1992. 160 p. ISBN 0–687–42656–1 BV15 .D6738 1992

Doran, Carol, and Thomas H. Troeger. *Trouble at the Table: Gathering the Tribes for Worship.* Lexington, Ky.: Lexington Volunteer Recording Unit, 2002. 5 audio cassettes.

Chapter 2, "Worship and Music: Hearing Again the Harmony of the Spheres," addresses cultural differences, the theological importance of music in the church, ways to recover music as prayer, religious ambivalence about music, the function of music in worship, the role of the church musician, the relationship between musician and pastor, and music as a pastoral art. Documented with endnotes. Also published on sound cassettes.

634. Egan, Raymond. "Music for Ritual: A Philosophy for the Practice of Church Music at the End of the Twentieth Century." D.M.A. dissertation. Los Angeles: University of Southern California, 1996. 71 p.

Dissertation consists of an essay, an original poem, and two original music compositions. Essay contends: "1) that church musicians need to compose their own music; 2) that they need to be fluent in a large number of musical vocabularies of their time and of previous times; and 3) that their work needs to be informed by a spiritual component." The poem and compositions illustrate the essay's main points.

635. Foley, Edward. *Music in Ritual: A Pre-Theological Investigation.* Washington, D.C.: Pastoral Press, 1984. 30 p. ISBN 0–912405–09–0 ML3000 .F64 1984

Essay originally presented at the annual meeting of the North American Academy of Liturgy in 1982. Raises the question, "Why is music integral to worship?" Addresses the question through discourse of music as powerful, music as communication, music as language, music as symbol, and music as ritual. Documented with endnotes.

636. Gebauer, Victor E. "Problems in the History of American Church Music." *The Hymn: A Journal of Congregational Song,* 41/4 (Oct. 1990): 45–48

Purpose: "to (I) review the character of writing about American church music history, (II) propose a new scheme for writing church music history and (III) clarify a few of the most basic points of tension between American music history and American church music history." Documented with endnotes.

637. Hoffman, Lawrence A., and Janet Roland Walton, eds. *Sacred Sound and Social Change: Liturgical Music in Jewish and Christian Experience.* Ind.: University of Notre Dame Press, 1992. vi, 352 p. ISBN 0–268–01745-X ML2900 .S2 1992

Originated from a conference held in 1986 sponsored by the School of Sacred Music at Hebrew Union College-Jewish Institute of Religion, Union Theological Seminary, and the Institute for Sacred Music at Yale University. Four sections: (1) "Reconstructing the Past: Sacred Sound from the Bible to Reform" explores past social changes and traces the development of the "sacred sound" in synagogues and churches; (2) "Exploring the Present: Sacred Sound in North America Today" examines the state of current sacred music in U.S. synagogues and churches; (3) "Composing Sacred Sounds: Four New Settings of Psalm 136" presents four settings of the Psalm, each reflecting a different musical tradition (Catholic, Methodist, Jewish, and Episcopal/Anglican); and (4) "Critiquing Sacred Sound: Perspectives on the Sacred and the Secular" comments on the state of musical composition in U.S. synagogues and churches. Essays most noteworthy on the topic of U.S. church music include: "North American Culture and Its Challenges to Sacred Sound" (Walton); "Catholic Prophetic

Sound after Vatican II" (M. T. Winter); "Present Stress and Current Problems: Music and Reformed Churches" (H. T. Allen, Jr.); "The Hymnal as an Index of Musical Change in Reform Synagogues" (B-E. Schiller); "Sacred Music in a Secular Age" (S. Adler); "'Sing a New Song': A Petition for a Visionary Black Hymnody" (J. M. Spencer); "Enculturation, Style, and the Sacred-Secular Debate" (V. C. Funk); and "On Swimming Holes, Sound Pools, and Expanding Canons" (Hoffman). Musical examples; expansive index.

638. Hughes, Ray. *Sound of Heaven, Symphony of Earth.* Charlotte, N.C.: MorningStar, 2000. 157 p. ISBN 1–878327–93–3 BV290 .H8 2000

According to the author, "It is not a book about improving musical techniques, developing your song-writing skills, or preparing for a Sunday morning worship service." Considers church music from a theological perspective.

639. Hustad, Donald P. *Jubilate II: Church Music in Worship and Renewal.* Carol Stream, Ill.: Hope Publishing, 1993. xxii, 599 p. ISBN 0–916642–17–8 ML3100 .H88 1993

Revision of the author's *Jubilate!: Church Music in the Evangelical Tradition* (Hope Publishing, 1981). Examines the musical practices of nonliturgical churches. Many sections relate to church music in the United States; two chapters are specifically devoted to the topic: "Music in Worship and Renewal in America through the 19th Century" and "Music in Worship and Renewal in America in the 20th Century." Musical examples and tables; bibliography of approximately 350 writings; expansive index.

640. Hustad, Donald P. *True Worship: Reclaiming the Wonder & Majesty.* Wheaton, Ill.: Shaw; Carol Stream, Ill.: Hope Publishing, 1998. 308 p. ISBN 0–87788–838–8 ML3100 .H89 1998

Appraises musical practices within nonliturgical worship traditions, specifically evangelical Protestant churches. Bibliography of nearly 130 writings; expansive index.

641. Johansson, Calvin M. "Some Theological Considerations Foundational to a Philosophy of Church Music." D.M.A. dissertation. Fort Worth, Tex.: Southwestern Baptist Theological Seminary, 1974. iii, 393 p.

Advocates seven viewpoints on church music, specifically addressing the church music scene in the United States: (1) church music must be creative; (2) must correspond to people's abilities, cultural environment, and insight; (3) must be correlative to the gospel; (4) must take into account the congregation's musical profile when choosing music; (5) should be well known or readily apprehensible; (6) should exhibit a balance between reason and emotion; and (7) must proceed in a manner which shows the faith action required of the Christian life. Bibliography of **nearly** six hundred writings.

642. Leaver, Robin A., ed. *Church Music: The Future: Creative Leadership for the Year 2000 and Beyond.* Princeton, N.J.: Westminster Choir College, 1990. 70 p. BV290 .C58 1989

Papers read at a symposium on church music sponsored by Westminster Choir College, October 15–17, 1989. The symposium was centered around five commissioned compositions: Samuel Adler's *Verses from Isaiah*, Ronald Arnatt's *New Songs of Celebration*, Richard Hillert's *Alleluia! Voices Raise*, Don Saliers' *Psalm 104*, and Richard Proulx's *Prelude and Introit: Open Wide the Windows of Our Spirits, O Lord.* Numerous contributors to the symposium. Papers address the future of church music, music for worship, instruments, musicians, and composing.

643. Leaver, Robin A., and Joyce Ann Zimmerman, eds. *Liturgy and Music: Lifetime Learning.* Collegeville, Minn.: Liturgical Press, 1998. x, 455 p. ISBN 0–8146–2501–0 MT88.L57 1998

Two parts: Part 1 covers worship and liturgy; Part 2 contains thirteen essays on liturgical music: "What is Liturgical Music?" (Leaver); "Liturgical Music as Music: The Contributions of the Human Sciences" (J. M. Joncas); "Liturgical Music: Its Forms and Functions" (R. F. Glover); "Liturgical Music as Liturgy" (W. T. Flynn); "Liturgical Music as Prayer" (K. Harmon); "Liturgical Music as Corporate Song 1: Hymnody in Reformation Churches" (Leaver); "Liturgical Music as Corporate Song 2: Problems of Hymnody in Catholic Worship" (F. C. Quinn); "Liturgical Music as Corporate Song 3: Opportunities for Hymnody in Catholic Worship" (M. J. Molloy); "Liturgical Music as Homily and Hermeneutic" (Leaver); "Liturgical Music, Culturally Tuned" (M. P. Bangert); "Liturgical Musical Formation" (D. E. Saliers); "Liturgical Music as Anamnesis" (Leaver); and "Liturgical Music: A Bibliographic Essay" (E. Foley). Latter essay presents a classified bibliography with discussion of more than four hundred writings. Essays documented with endnotes; no index.

644. McCalister, Lonnie. "Developing Aesthetic Standards for Choral Music in the Evangelical Church." D.M.A. dissertation. Norman: University of Oklahoma, 1987. vii, 166 p.

Describes the state of choral music in Evangelical churches and offers suggestions for the development of aesthetic standards. Provides classified list of ninety-one recommended choral works. Bibliography of approximately 160 writings.

645. Milligan, Thomas B., ed. *On the State of Church Music.* Jefferson City, Tenn.: Louis and Mary Charlotte Ball Institute of Church Music, and Center for Church Music, Carson-Newman College, 1993. vi, 53 p. ML3106 .O5 1993

Eight essays by prominent Protestant church musicians from the United States and England focusing on current church music issues within Baptist, Methodist, Lutheran, and Episcopal/Anglican traditions. Essays: "The Present State of Church Music" (S. Amerson); "Church Music 1992—Hither? Thither?" (L. Ball); "What's Ahead for the Church in the 1990's?" (H. M. Best); "*Lex Cantandi—Lex Credenti*" (W. Forbis); "What Price Tradition?: An Examination of What We are Really Singing in Our Churches" (A. Luff); "The Present State of Church Music: A Personal View" (N. H. Tredinnick); "The Present State of Church Music: Historical and Theological Reflections" (P. Westermeyer); and "The Reconstruction of USA Church Music Education" (C. R. Young).

646. *On the State of Church Music, IV: Worship: Unity or Diversity.* Jefferson City, Tenn.: Louis and Mary Charlotte Ball Institute of Church Music, and Center for Church Music, Carson-Newman College, 1996. 41 p. ML3106 .O5 1996

Four essays on the theme "unity or diversity." All contributors active in the United States. Essays: "Integrating Music and Preaching into the Service of Worship" (W. E. Hull); "Aesthetics, Unity, and Diversity in Church Music" (D. B. Austin); "Worship: Unity or Diversity" (W. L. Forbis); and "The Poetry of God" (Hull). Endnotes follow some essays; photos of contributors.

647. Pass, David B. *Music and the Church.* Nashville, Tenn.: Broadman Press, 1989. 131 p. ISBN 0–8054–6814–5 ML3001 .P18 1989

Outlines the author's theory regarding mission and use of music in the worship service. Tables and illustrations; documented with footnotes.

648. Pfatteicher, Philip H. *The School of the Church: Worship and Christian Formation.* Valley Forge, Pa.: Trinity Press International, 1995. ix, 149 p. ISBN 1–56338–110–9 BV178 .P45 1995

Chapter 4 of this study (sixteen pages) examines "liturgical language" through musical art. Documented with endnotes; general expansive index.

649. Pottie, Charles S. *A More Profound Alleluia!: Gelineau and Routley on Music in Christian Worship*. Washington, D.C.: Pastoral Press, 1984. viii, 104 p. ISBN 0–912405–12–0 ML3001 .P68 1984

Explores the writings of Joseph Gelineau (1920–) and Erik Routley (1917–1982). Representing two different perspectives, Gelineau embraces the Catholic liturgical tradition, and Routley, the Reformed, free church worship tradition. Bibliography of 127 writings by Gelineau and 123 by Routley.

650. Riedel, Johannes, ed. *Cantors at the Crossroads: Essays on Church Music in Honor of Walter E. Buszin.* St. Louis, Mo.: Concordia, 1967. xvii 238 p. ML3000.1 .C3

Twenty essays. Three of the essays relate to church and worship music in the United States: "Twentieth-Century Church Music—American Style" (E. Copes); "The Dilemma of Church Music Today" (H. E. Pfatteicher); and "The Struggle for Better Hymnody" (A. Haeussler). Musical examples; some essays documented with endnotes; classified, general bibliography of approximately five hundred writings.

651. Sanders, John B., and Robert C. Dvorak. *Music in the Church.* South Hamilton, Mass.: Sanders Christian Foundation, 1977. 5 audio cassettes.

A series of recorded lectures and musical presentations. Topics: "Music in Contemporary American Society;" "Music as an Art Form in the Church;" "Use of the Psalm Form in Worship;" "New Testament Influence on Church Music;" "Hymns, as We Know Them;" "Hymnic Contributions of the Roman Church;" "Contributions of Luther & the Reformers;" and "Representative Writers & Traditions."

652. Seel, Thomas Allen. *A Theology of Music for Worship Derived from the Book of Revelation.* Metuchen, N.J.: Scarecrow Press, 1995. xi, 209 p. ISBN 0–8108–2989–4 ML3001 .S394 1995

Stated purpose: to demonstrate that the New Testament, especially the Book of Revelation, has "practical, 'down-to-earth' and specific things to say regarding the use of music for worship in the life of the contemporary church." Revision of author's doctoral dissertation (Southern Baptist Theological Seminary, 1990). Several appendixes, including location of musical references in the Book of Revelation. Classified bibliography of more than two hundred writings; index.

653. Wienandt, Elwyn A., ed. *Opinions on Church Music: Comments and Reports from Four-and-a-Half Centuries.* Waco, Tex.: Baylor University Press, 1974 (rep. 1984). x, 214 p. ISBN 0–918954–30–4 (pbk.) ML3000 .W535

Forty-nine essays, letters, and memoirs in chronological order from the sixteenth century through middle of twentieth century. International in scope. Selections that address U.S. church music issues: (1) extract from *The Continental Harmony* (1794) by William Billings; (2) 1845 address to the American Musical Convention by Thomas Hastings; (3) address by Raymond Seely on the function of congregational singing; (4) "Motu Proprio of Pope Pius X on Sacred Music" (1903); (5) "A Survey of Music in America" by O. G. T. Sonneck (1913); (6) an excerpt on individualism drawn from Archibald T. Davison's *Protestant Church Music in America* (1933); and other mid- to late-twentieth-century contributions by H. C. Colles, James F. White, Howard D. McKinney, Frank Cunkle, Dave Brubeck, and Stephen Koch. Bibliography of about sixty writings; expansive general index.

See also: Wicker, Vernon, ed. *The Hymnology Annual: An International Forum on the Hymn and Worship* (item 450)

A CAPPELLA

654. Johnson, Aubrey. *Music Matters in the Lord's Church.* Nashville, Tenn.: 20th Century Christian, 1995. 140 p. ISBN 0–89098–141–8

Advocates the use of a cappella singing only within the worship service. Bibliography of thirty writings.

CONTEMPORARY MUSIC AND WORSHIP

655. Baker, Wesley L. "Worship, Contemporary Christian Music, and Generation 'Y'." D.Min. dissertation. Due West, S.C.: Erskine Theological Seminary, 2000. xii, 211 p.

Purpose: "addresses the church's exploration of the effectiveness of Contemporary Christian Music in worship for engaging young people, 'Generation Y,' with the Christian Gospel." Provides a brief history of church music from Old Testament references through twentieth-century church music practices in the United States. Numerous tables; bibliography of approximately fifty-five writings and Web sites.

656. Best, Harold M. *Music through the Eyes of Faith.* San Francisco: HarperSanFrancisco, 1993. xii, 225 p. ISBN 0–06–060862–5 ML3871 .B47 1993

Theological approach to musical practice. Examines the issue of musical quality and discusses the incorporation of contemporary Christian music in the worship service. Bibliography of sixty-seven writings; expansive index.

657. Cloud, David W. *Contemporary Christian Music under the Spotlight.* Port Huron, Mich.: Way of Life Literature, 1998. 489 p. ISBN 1–58318–057–5 ML3187.5 .C6248

Argues against the use of contemporary Christian music in worship services. Includes a 241-page directory of contemporary Christian musicians, often with disparaging comments. Also examines "worldly" aspects in southern gospel music. Annotated bibliography of approximately thirty writings and a list of resources for music scores and sound recordings; classified bibliography (miscellaneous; contemporary Christian music; sacred music; song leading; and southern gospel) of nearly two hundred writings; no index.

658. Fisher, Tim. *The Battle for Christian Music.* Foreword by John Vaughn. Greenville, S.C.: Sacred Music Services, 1992. xv, 211 p. ML3111 .F5

Described by the author as an "attempt to defend the historical position practiced and held throughout church history." Focuses on the use of contemporary Christian music in worship services. Bibliography of forty-seven writings; scripture index.

659. Frame, John M. *Contemporary Worship Music: A Biblical Defense.* Phillipsburg, N.J.: P & R Publishing, 1997. xii, 212 p. ISBN 0–87552–212–2 ML3187.5 .F73 1997

Theological approach to issues surrounding the use of contemporary worship music in Christian services. Documented with endnotes; recommended listing of approximately 150 songs for use in worship service; three indexes: scripture references, song titles, and persons.

660. Howard, Jay R., and John M. Streck. *Apostles of Rock: The Splintered World of Contemporary Christian Music.* Lexington: University Press of Kentucky, 1999. viii, 299 p. ISBN 0–8131–2105–1 ML3187.5 .H68 1999

Draws upon H. Richard Niebuhr's work in *Christ and Culture* (Harper, 1951; Faber and Faber, 1952). Classifies contemporary Christian music (CCM) into three categories: "Separational CCM," "Integrational CCM," and "Transformational CCM." Examines criticism often associated with the contemporary Christian music business. Photographs; discography of nearly 160 recordings; bibliography of approximately two hundred writings; expansive index.

661. Lucarini, Dan. *Why I Left the Contemporary Christian Music Movement: Confessions of a Former Worship Leader.* Webster, N.Y.: Evangelical Press, 2002. 141 p. ISBN 0–85234–517–8 ML420.L83 L83 2002

Not available for review. "Lucarini, now a businessman, was a worship leader for several evangelical churches and was also a rock music performer, arranger and composer. Writing from his own experience, Lucarini questions the use of contemporary music in worship" (publisher).

662. Lynch, Ken. *Biblical Music in a Contemporary World.* Foreword by Phil Gingery. Chester, Pa.: Ken Lynch, 1999. 130 p. ISBN 0–7392–0394–0 ML3187.5 .L96 1999

Takes a stand against the use of contemporary Christian music in worship. Illustrations; bibliography of seventeen writings and a recommended list of fifteen writings.

663. Makujina, John. *Measuring the Music: Another Look at the Contemporary Christian Music Debate.* Foreword by Calvin M. Johansson. Salem, Ohio: Schmul Publishing, 2000. 303 p. ISBN 0–88019–403–0 ML3187.5 .M35 2000

Argues against the use of contemporary Christian music in worship. Illustrations; bibliography of nearly five hundred writings.

664. Mitchell, Robert H. *I Don't Like That Music.* Foreword by Donald P. Hustad. Carol Stream, Ill.: Hope Publishing, 1993. 142 p. ISBN 0–916642–49–6 BV290 .M58 1993

Addresses the debate over the inclusion of popular, ethnic, and "classical" music in worship. Purpose: "to suggest that there are good reasons why one should attempt to understand, accept and enter into a variety of kinds of worship." Documented with endnotes.

665. Payton, Leonard R. *Reforming Our Worship Music.* Wheaton, Ill.: Crossway Books, 1999. 48 p. ISBN 1–58134–051–6 ML3001 .P23 1999

Deals with the controversy over the use of contemporary sacred music in Christian worship services. Advocates the use of music with rhythms determined by text rather than dance rhythms, musical training for pastors and children, investing in congregational accompaniment, emphasis on efficacious music and avoidance of commercially-produced music, and more singing by congregation. Bibliography of ten writings.

666. Peters, Dan, Steve Peters, and Cher Merrill. *What about Christian Rock?* Minneapolis, Minn.: Bethany House, 1986. 223 p. + 1 audio cassette. ISBN 0–87123–672–9 ML3187.5 .P47 1986

Addresses the use of Christian rock in worship. Incorporates statements by performers. Photographs; brief lists of writings, audio tapes, videos and films, and festivals; index. Audio cassette provides interviews with performers and musical selections.

667. Resch, Barbara J. "Adolescents' Attitudes toward the Appropriateness of Religious Music." D.Mus.Ed. dissertation. Bloomington: Indiana University, 1996. x, 152 p.

Adolescent subjects drawn from Massachusetts, Indiana, Tennessee, and California listened to recordings "of 40 musical excerpts representing current American church music practice and indicated their perceived level of appropriateness of each example." High ratings were scored for traditional "classical" choral and instrumental music; low ratings were scored for religious rock music. Scorings reveal that the adolescent subjects' attitudes toward musical appropriateness are not the same as their musical preference. Statistical tables; bibliography of more than 150 writings.

668. Riccitelli, James Michael. *Sing a New Song: When Music Divides the Church.* Blissfield, Mich.: H & E Berk, 1997. 175 p. ISBN 0–9658900–0–7 BV290 .R53 1997

Discussion of the impact of secular musical styles, specifically rock music, on the Christian church service. Offers twelve general principles regarding the use or exclusion of secular music in worship. Bibliography of approximately 120 writings.

669. Sears, Gordon E. *Is Today's Christian Music "Sacred"?* Foreword by Harold DeCou; conclusion by Rudy Atwood. Coldwater, Mich.: Gordon E. Sears, n.d. (ca. 1990–1992). 32 p. ML3187.5 .S43

Calls into question "the validity of Christian contemporary music in light of the Word of God." Argues against use of contemporary Christian music in worship services.

670. Seidel, Leonard J. *Face the Music: Contemporary Church Music on Trial.* Springfield, Va.: Grace Unlimited Publications, 1988. xv, 163 p. ISBN 1–703–644–1468 ML3111 .S44 1988

Advocates that the Christian church should not incorporate rock music in the worship service. Bibliography of more than sixty writings.

671. Wheaton, Jack. *The Crisis in Christian Music.* Oklahoma City, Okla.: Hearthstone, 2000. 184 p. ISBN 1–57558–062–4 ML3187.5 .W44 2000

Crisis defined as "a weakness in bringing contemporary Christian music . . . into the sanctuary as acceptable praise music for a worship service." Recommends acceptable music alternatives. Provides suggestions for arranging for praise bands and modern choirs. Offers a list of recommended praise songs. Bibliography of almost sixty writings; discography and videography of nearly thirty items; no index.

672. Wohlgemuth, Paul W. *Rethinking Church Music.* Foreword by Don G. Fontana. Rev. ed. Carol Stream, Ill.: Hope Publishing Company, 1981. x, 101 p. ISBN 0–916642–15–1 ML3000 .W7

Meditation on church music, primarily drawing on the author's extensive experience as a music minister and church music academician. No attempt to place trends in contemporary church music into historical perspective. Makes the point that church music has repeatedly adapted to change in church and society and that tension between the traditional and the new has been a constant in church history. Approaches the question of contemporary gospel music and Christian rock music from a conservative Evangelical viewpoint. Briefly considers issues and concerns that impinge on church music programs. Contains little that would be new to an experienced church musician, but could serve to educate pastors and laymen to the problems, contradictions, and pressures facing church music programs.

673. York, Terry W. *America's Worship Wars.* Peabody, Mass.: Hendrickson, 2003. xviii, 138 p. ISBN 1–56563–490-X BV8 .Y67 2003

Music has been a key element in the conflicts over worship styles. Beginning with the worship music situation in the 1960s, the author offers "a mixture of documented history, observation and interpretation. . . . It is one person's intense but limited commentary. Its documented history is true and its observation and interpretation are faithfully recounted and presented." York explores the perspectives, agendas and tactics of those involved in the conflicts, traces the recent history of church music, explores the connection with America's "culture wars," and suggests some theological perspectives. Bibliography of more than eighty writings and sixteen hymnals, song collections, and musicals; index.

See also: Price, Milburn. "The Impact of Popular Culture on Congregational Song" (item 290)

INSTRUMENTAL MUSIC

674. Curl, Charles Edward. "Developing a Theology of Worship to Inform the Use of Musical Instruments in Local Church Worship." D.Min. dissertation. Madison, N.J.: Drew University, 1981. v, 123 p.

Reviews the usage of instruments in worship from the early church to the present; develops a theory defining possible "biblical criteria" for the use of instruments, with greatest focus on the organ. Five statements as guidelines for determining the appropriate use of instrumental music: (1) "Because musical instruments were frequently used in the Old Testament days in worship, and because their use is not prohibited in the New Testament, various instruments of music may be used in our worship today;" (2) "Music in general, and singing in particular, are important to contemporary worship;" (3) "The prophets call us to be ever sure that what we do at worship, and what we do in the rest of our lives, are in harmony and basic agreement;" (4) "Edification as taught by the apostle Paul is the ruling guideline for judging worship practice. No practice is acceptable if it does not contribute to the spiritual building up of the total group;" and (5) "Inward spiritual attitudes are always more important than any outward form used in worship." Classified bibliography of approximately 120 writings.

675. Girardeau, John L. *Instrumental Music in the Public Worship of the Church.* Richmond, Va.: Whittet & Shepperson, 1888. 208 p. ML3001 .G51; reprint, Haverton, Pa.: New Covenant Publication Society, 1983. ML3001 .G51 1983; reprint, Edmonton, Alberta, Canada: Still Waters Revivals Books, 1992. 208, 4 p. ML3001 .G51 1992

A nineteenth-century perspective regarding the "unjustifiable employment of instrumental music in public worship." Organized into six parts: general argument from scripture; argument from the Old Testament; argument from the New Testament; argument from the Presbyterian standards; historical argument; and arguments in favor of instrumental music considered. Documented with endnotes.

676. *Instrumental Music: Faith or Opinion?* Huntsville, Ala.: Publishing Designs, 1991. xii, 186 p. ISBN 0–929540–10–7 ML3001 .F74 1991

Papers read at Freed-Hardeman University, October 12, 1991. Addresses the issue of instrumental music in worship services. Includes transcriptions of question-answer sessions. Contributors: L. James, C. May, Jr., B. Swetmon, and D. Lipe.

X

Church and Worship Music Web Sites

GENERAL WORKS

677. Christian Classics Ethereal Library (CCEL).

http://www.ccel.org/

Describes itself as a selection of "classic Christian books in electronic format." Consists of more than books; also includes various indexes and MIDI files. A number of resources relate to hymns and hymnody. A sampling of resources relating to music in the United States are:

Borthwick, Jane. *Hymns from The Land of Luther.*

http://www.ccel.org/ccel/borthwick/hll.html

Web resource under construction. Electronic reprint of the 1866 edition of the text-only hymnal.

The Christian Classics Ethereal Library Hymn Tune Archive.

http://www.ccel.org/cceh/

"A database and indexed archive of public-domain hymn tunes and chants in electronic formats including MIDI files, printable sheet music, and editable electronic musical scores. . . . It does not include hymn texts or lyrics."

The Hymnal.

http://www.ccel.org/ccel/anonymous/eh1916.html

Web resource under construction. Electronic reprint of the 1920 edition of the text-only hymnal of the Protestant Episcopal Church in the USA.

Plantinga, Harry. *The Online Southern Harmony.*

http://www.ccel.org/s/southern_harmony/

Hymns from *The Southern Harmony and Music Companion* by William Walker indexed under four headings: Tunes; First Line; Composer; and Meter. Links provided to twenty-five of the most popular hymns. Includes information about the collection and its compiler.

Watts, Isaac. *Divine and Moral Songs for Children.*

http://www.ccel.org/ccel/watts/divsongs.html

Web resource under construction. Electronic reprint of the 1866 edition of the text-only hymnal.

Wesley, John. *A Collection of Hymns, for the Use of the People Called Methodists.*

http://www.ccel.org/w/wesley/hymn/jw.html

Easily searchable electronic version of the 1889 edition of the text-only hymnal.

678. *ChristSites.*

http://www.christsites.com/

International in scope. Twelve main topics, including "Arts & Media" and "Publishing." "Arts & Media: Music" lists more than 1,700 Web sites under the subheadings: Artists; Audio; Choirs; Concert Listings; Directories; Distribution; DJs; Genres; Labels; Magazines; Online Resources; Recording; and Reviews.

679. *The Church Music Report* (TCMR).

http://www.tcmr.com/links.html

Web journal. Provides extensive list of links to sites of interest to church musicians. Some of the links are: children's music; church Web sites; clinicians/consultants; composers/arrangers; discount music dealers; handbell/supplies/music; keyboards; music conferences; music dealers; multimedia; music education; music organizations; orchestra; professional organizations; search engines; southern gospel music; vocal training; and worship resources.

680. Dolan, Marian. *TheoTech: An Internet Primer for Pastors, Musicians and Worship Planners.*

http://www.pitts.emory.edu/theotech/

Lists selected Web links under ten topics: Mega-sites; Choral; Copyright; Hymnals (online, denominational, hymn resources); Multi-Cultural; Notation Software; Organizations & Guilds; and Psalmody.

681. *emusicquest.*

http://www.emusicquest.com/

The *Music in Print* series (items 85 and 120) in a searchable online database. Access by subscription only. Lists music in-print for the following categories: sacred and secular choral; organ; classical vocal; orchestral; string; classical guitar; woodwind; piano; and miscellaneous (which includes band, brass, handbell, harp, percussion, etc.).

682. Inoue, Tadashi. *Church Music Research Tools.*

http://homepage3.nifty.com/dikaios/CMstudy/English/title.html

English version of the original Japanese version of Inoue's extensive list of church music resources. Partially annotated. Site "designed to provide bibliographic resources for studying church music in its historical sense." Some resources relate to Japanese church music, whereas other resources are international in scope, including resources about U.S. church music.

683. *Internet Theology Sources: Liturgical Studies and Liturgical Music.*

http://www.csbsju.edu/library/internet/theoltgy.html

Lists twenty-one Web links to varied liturgical music sources. Includes: Choral Archives; Church Music & the Kinetic Arts; Classical MIDI Organ Stop; Southern Harmony; etc. Maintained by the Clemens Library of the College of Saint Benedict, Minnesota, and the Alcuin Library of Saint John's University, Collegeville, Minnesota.

684. Lyrics.com.au

http://www.lyrics.com.au

Although an Australian-produced Web site, provides links to many church music Web sites relating to the United States, including sites on hymn accompaniment, hymnal indexes, hymnody, psalters, and shape note music. Many of the links relate specifically to lyrics, but not all; also includes bibliographies and historical information.

685. Murrow, Rodney C. *The Little House on the Internet.*

http://www.pldi.net/~murrows/

One section of this Web site is titled "Resources for Church Musicians: An Online Workshop." Described by the author as "a meta-list of resources useful for musicians, pastors, and worship leaders." Comprised of bibliographies of music, bibliographies of writings, and other resources relating to a variety of topics, including organ, piano, choral music, handbells, instrumental music, hymnody, MIDI and technology, philosophy of church music, publishers, and retailers.

686. National Association of Church Musicians (NACM).

http://www.nacmhq.org/content/home/

Largest nondenominational association of church musicians and other worship leaders in North America for more than sixty years. Produces *The Journal*, the latest issue of which can be downloaded. Annual Convention Brochure available on the site.

687. *The Pentecostal Resource Center: Subject Research Guides & Links.*

http://library.leeuniversity.edu/guides/

Published by William G. Squires Library, Lee University, and the Dixon Pentecostal Resource Center, Church of God Theological Seminary, both located in Cleveland, Tennessee. Covers a variety of topics, including "Arts and Music." Under the category "Music" are six subcategories: Careers, Church Music, Instrumental Music, Music Analysis and Biography, Music Education, and Music Journals. Within the subcategory of "Church Music" are general church music resources (including music databases, general church music and music ministry, hymnology and liturgy, instrumental music, and vocal music), "Modern Genres," and "Special Categories." The "Modern Genres" section is especially useful for Web links related to southern gospel and Christian country, spirituals and black gospel, contemporary Christian music, and Christian rock and Christian alternative (e.g., Celtic Christian tunes).

688. *Questia: The World's Largest Online Library.*

http://www.questia.com/

A subscription service ($120/year): Full text available of nearly fifty thousand books and almost four hundred thousand journal, magazine, and newspaper articles. A search for "church music" resulted in just fewer than fourteen hundred hits, most of which were books, although nearly fifty journal and almost ninety magazine articles were listed. Service probably best used for research.

689. Royal School of Church Music in America (RSCM America).

http://www.rscmamerica.org/

Founded 1927; RSCM America is now the largest organization promoting the use of music in Christian worship (eleven thousand–plus members worldwide). Seeks to find the best in all styles and periods of church music. Cardinal principle: Whatever music is used should be best of its kind. Web site includes information on membership, upcoming events, newsletter, and so on.

690. Sibelius Academy.

http://www2.siba.fi/Kulttuuripalvelut/music.html

Sibelius Academy is a conservatory in Helsinki, Finland that "provides a list of music sites that would rival a Vienna guidebook" (*PC Magazine*). Provides Web links to jazz, blues, rock, popular music, famous composers, gospel, instruments, research, theory, opera, etc. The "Church Music and Other Christian Music" section offers links to twenty-eight international Web sites.

RELIGIOUS AND ETHNIC GROUPS

African American

691. Center for Black Music Research (CBMR).

http://www.cbmr.org/

CBMR is a research unit of Columbia College, Chicago, Illinois. CBMR documents, collects, preserves, and disseminates information about black music worldwide. Supports study in African-American sacred folk music and gospel music, among other black music genres.

Armenian

692. Diocese of the Armenian Church of America (Eastern). *The Armenian Church.*

http://www.armenianchurch.org/worship/music/music.html

Discusses sacred music in the Armenian church, in addition to information on worship, services, sermons, and so on.

Baptist

693. Fellowship of American Baptist Musicians (FABM).

http://www.fabm.com/

Provides information on the association, the annual Conference for Church Music, job openings, the Lending Library, etc.

Catholic

694. Church Music Association of America (CMAA).

http://www.musicasacra.com

Established in 1964 through the merger of the American Society of St. Cecilia (est. 1874) and the St. Gregory Society (est. 1913). Web site provides information about the CMAA organization, educational programs, the Society's journal *Sacred Music* (item 146), and a few publications of the Catholic Church Music Associates.

695. National Association of Pastoral Musicians (NPM).

http://www.npm.org/

The NPM is "primarily composed of musicians, musician-liturgists, clergy, and other leaders of prayer devoted to serving the life and mission of the Church through fostering the art of musical liturgy in Catholic worshiping communities in the United States of America." Publishes *Pastoral Music* (item 144) and *The Liturgical Singer*, a quarterly periodical for "Cantors, Choir and Ensemble Singers," which provides practical help regarding vocal techniques and the singer's role in the liturgy. The NPM Web site offers information regarding music education (Divisions: Music Education) and an extensive bibliography of choral anthems organized by church year (Resources: Choral Anthems).

Christian Reformed

See: *Reformed Worship: Resources for Planning and Leading Worship* (item 145)

Church of God

696. *Music for the Church of God.*

http://www.cgmusic.com/

Includes news, information on hymnals and psalters, a library, Web links, and so on.

Episcopal/Anglican

697. Association of Anglican Musicians (AAM).

http://www.anglicanmusicians.org

Describes itself as "an organization of musicians and clergy in the Episcopal Church and throughout the Anglican communion." Web site includes information on the organization and lists several monographic publications, for example, *Bibliography of Useful Resources for the Church Musician* (M. A. Neilson, A. C. Reed).

Evangelical Lutheran

698. Evangelical Lutheran Church in America (ELCA).

http://www.elca.org/

A search for the word "music" results in a list of more than two thousand ELCA and other church music documents available online.

699. *Lift Up Your Hearts: Worship & Spirituality Site of the Evangelical Lutheran Church in Canada* (ELCIC).

http://www.worship.ca/

Provides Web links to thousands of international Web sites, including many U.S. sites. Section 7, "Church Music & the Kinetic Arts," alone includes more than three hundred links under twelve subheadings: associations, guilds, societies; journals; bells; chant; hymns and liturgical music; organ; voice/choir; music studies & scholarship; music publishers; music link collections; other music; and kinetic arts.

Greek Orthodox Archdiocese

700. Greek Orthodox Archdiocese of America.

http://www.goarch.org/en/multimedia/

Under "Multimedia Programs" are pages for audio, video, virtual reality, live broadcasts, etc. The audio pages, for example, offer samples of "Hymns from the Liturgical Tradition of the Orthodox Christian Church." Included are hymns for the major feasts of the church, Orthros hymns, and samples from the divine liturgy.

Jewish

701. The Zamir Choral Foundation. *Links of Interest.*

http://www.zamirfdn.org/links.shtml

Links to Web sites related to Jewish choral music in the United States. Sections include: foundations; U.S. based professional Jewish choirs; international Jewish choral groups; university/seminar programs; and a few general choral links/resources.

Lutheran

702. Association of Lutheran Church Musicians (ALCM).

http://www.alcm.org/alcm/html/index.jsp

Publishes *Cross Accent: Journal of the Association of Lutheran Church Musicians.* Web site describes the purpose of the organization and includes sections on publications, job placement, recordings, and so on.

703. The Lutheran Church Missouri Synod (LCMS).

http://www.lcms.org

Articles on aspects of music in Lutheran churches, worship resources list, event calendar, Web links, and so on.

Orthodox

704. Pan-Orthodox Society for the Advancement of Liturgical Music (PSALM).

http://www.orthodoxpsalm.org/

PSALM is "dedicated to the advancement, excellence and growth of traditions in Orthodox liturgical music and chant, with a special missionary emphasis on fostering these traditions wherever English is used as a liturgical language." Publishes music online, presents news and lists events, includes job notices, and so on.

Presbyterian

705. Presbyterian Association of Musicians (PAM).

http://horeb.pcusa.org/pam/

Lists news and events, conference information, job listings, professional concerns, certification requirements, and so on.

Seventh-Day Adventist

706. *Seventh-Day Adventist Digital Hymnal.*

http://www.tagnet.org/digitalhymnal/

Collection of MIDI files, lyrics, and historical notes about authors and composers of hymns, based on *Seventh-Day Adventist Hymnal.* Lyrics in English and Spanish. Many well-known hymns included. Provides Web links to MIDI digital hymnals (including HymnSite.com, The Cyberhymnal, and others) and other resources (e.g., Christian MIDI Collections). Created and maintained by Pablo A. Aguilar. Not officially sponsored by the Seventh-Day Adventist Church.

United Methodists

707. The Fellowship of United Methodists in Music and Worship Arts (FUMMWA).

http://members.aol.com/fummwa/fummwa.htm

"A professional agency of the United Methodist Church that serves as a resourcing organization to its members. It provides training on all aspects of worship. . . . FUMMWA is fellowship of those involved in worship ministries, serving: Pastors, Worship Leaders, Dancers, Musicians, Dramatists, Visual Artists, Christian Educators." Publishes the bimonthly journal *Worship Arts.*

CHORAL MUSIC

708. American Choral Directors Association (ACDA).

http://www.acdaonline.org/

Publishes *The Choral Journal* (item 137), the leading choral music periodical in the United States. The journal includes dozens of choral music reviews each year, including many reviews of sacred works.

709. *The Choral Public Domain Library* (CPDL).

http://cpdl.org/

A free sheet music archive. More than six thousand scores, many of them of sacred works, are available for download.

710. *ChoralNet: The Internet Center for Choral Music.*

http://www.choralnet.org/resources/displayResources.phtml?category=8

Church music resources are listed under the headings: church music issues; denominational music; hymns and hymnals; and church music links.

711. *Choristers Guild.*

http://www.choristersguild.org

"Choristers Guild . . . enables leaders to nurture the spiritual and musical growth of children and youth." Information on resources for directors of children's, youth, and handbell choirs. Publishes music for children's and youth groups. Entire catalog may be searched online. Publishes *The Chorister* (item 138).

CONTEMPORARY CHRISTIAN MUSIC

712. Arthurs, Keith. *The Omnilist of Christian Links.*

http://www.arthursfamily.com/omnilist/

Links to hundreds of contemporary Christian Web sites. Sites are ranked as silver or gold. "Music" section lists links under: Music Reviews, E-Zines, Record Companies, Music Charts/Playlists, FAQs, Festivals, Hymns and MIDI sites, etc. "Musicians" section lists numerous contemporary Christian performers and composers.

713. CCM Magazine.com

http://www.ccmcom.com/

News and information for contemporary Christian performers and fans. Publishes the monthly magazine *CCM Magazine: Faith in the Spotlight* (see 136).

714. *CCMusic: Contemporary Christian Music.*

http://www.ccmusic.org/

Hosted by the Christian Music Place (item 715). Provides a discussion zone, artists list, music directory, "CCMusic Notes," and advertising information. "CCMusic Info" offers information about CCMusic and free Christian newsletters.

715. Christian Music Place.

http://www.christianmusic.org/cmp

Provides information on Christian artists, albums, music festivals, recording and distribution of Christian music, touring, church music, ministry opportunities, and so on. Publishes the online journal *CCMusic: Contemporary Christian Music* (see 714).

DISCOGRAPHIES

716. Moore, Berkley L. *Recordings Index.*

http://www.members.aol.com/berkmoore/Recordings

A discography of recordings of *Sacred Harp* music organized by title and by album title. Includes albums recorded at shape note singings, other recordings by shape note singers, shape note styled recordings by other singers, arrangements of shape note music, village carols, and West Gallery songs.

GOSPEL MUSIC

717. *American Gospel Music Directory.*

http://www.americangospel.com/

Directory lists gospel solo artists and groups, gospel songwriters, one international gospel link, and business related to gospel music.

718. *The Cyber Hymnal.*

http://www.cyberhymnal.org/

Includes more than forty-six hundred Christian hymns and gospel songs from many denominations. The resource provided as a public service.

719. *The Gospel Music Archive.*

http://www.gospel.boltblue.net/

Guitar chords and lyrics for hundreds of traditional and contemporary gospel songs. Information on other guitar-related matters.

HYMNS AND HYMNODY

720. Ehymnal.com.

http://ehymnal.com/

Includes pages entitled Hymns Online, Interactive Hymns, Hymns in Praise Style, Hymnspirations, Devotionals, etc. Hymns Online allows the user to click on hymn title or first line and locate scores in PDF format, MIDI files, Windows Media, Images, NWC files, interactive hymns, or devotionals. Not all formats available under all hymns.

721. Fasola.org.

http://fasola.org/

Information and Web links to the tradition of shape note or *Sacred Harp* singing. Also includes information about the *New Harp of Columbia*, *Southern Harmony*, *Christian Harmony* traditions, as well as the *West*

Gallery tradition from the British Isles. Provides links to articles on the practice, resource guides, maps with directions to rural singings, a list of singing groups, annual singings, and so on. Includes Chris Thorman's "The Sacred Harp Online Index" which indexes the lyrics to all the songs in the 1991 edition of *The Sacred Harp* searchable by title, page number, first line, composer, poet, meter, and key word. Cross referenced with Robert Stoddard's *Northern Harmony Online Index* (item 726). The "Shape Note Bibliography" by John Bealle is an impressive bibliography on all aspects of the shape note tradition. Resources are listed and defined under the terms: concepts, culture groups, influences, musical forms, religious groups, tunebook compilers, tunebooks, and geographic regions.

722. The Hymn Society of the United States and Canada.

http://www.thehymnsociety.org/home.html

Publishes *The Hymn: A Journal of Congregational Song* (item 142), a quarterly journal containing both practical and scholarly articles, and *The Stanza*, a semiannual newsletter. Web site includes information on the annual conference, workshops, tours, new hymns, and so on.

723. *The Hymn Tune Index.*

http://hti.music.uiuc.edu/

The index database "contains all hymn tunes printed anywhere in the world with English-language texts up to 1820, and their publication history up to that date." May be searched by tune, text, source, or composer. Supplemental information includes history, scope, Web links, and so on.

724. Sabol, Steven L. *Sacred Harp and Related Shape-Note Music: Resources.*

http://www.mcsr.olemiss.edu/~mudws/resource/

Resource topics include: annotated bibliographies of tunebooks, hymnals, music books, and internet resources; annotated discographies of recent recordings and older recordings still available; annotated videographies of videotapes and DVDs; descriptions of singing schools and camps; and so on.

725. *Sacred Harp Singing in Texas.*

http://www.texasfasola.org/

Privately owned Web site maintained by Gaylon L. Powell. Presents historical background of shaped note singing. Supplies biographical sketches of eight composers and poets whose works appear in *The Sacred Harp.* Provides links to other *Sacred Harp* Web sites.

726. Stoddard, Robert. *Northern Harmony Online Index.*

http://stoddardfamily.home.comcast.net/0Index.html

Indexes the lyrics to all the songs in the 4th edition of *Northern Harmony: Plain Tunes, Fuging Tunes and Anthems from the Early and Contemporary New England Singing Traditions* edited by Larry Gordon and Anthony G. Barrand. Indexed by title, page number, first line, composer, poet, meter, and key word. Cross referenced with Chris Thorman's "The Sacred Harp Online Index" (item 721).

See also: *The Cyber Hymnal* (item 718)

ORGAN MUSIC

727. American Guild of Organists (AGO).

http://www.agohq.org/home.html

Publishes *The American Organist* (item 134). The guild is a national association serving the organ and choral music fields.

728. Organ Historical Society (OHS).

http://www.organsociety.org/

Provides information about the OHS organization, membership, historic organs, and OHS organ tours. The "On-Line Catalog" lists recordings, books, videos, and classical sheet music related to the pipe organ and its music. Publishes *The Tracker: Journal of the Organ Historical Society* (item 149).

PUBLISHERS

729. Church Music Publishers Association (CMPA).

http://www.cmpamusic.org/html/main.isx

An organization of publishers of Christian music "which has a strong spiritual dimension." Provides information on nearly fifty publishers. List of Web links included.

Subject Index

Numerals are filed before letters, not as though they were spelled out. Initial articles in all languages are ignored in filing. Hyphenated words are considered as separate words in filing. Citations provide annotation number.

B

C

D

E

F

G

H

I

J

K

L

M

N

O

P

Q

R

S

T

U

V

W

Y

Z

Author Index

Entries are given for all authors, including joint authors, contributors, and editors. Names of translators are excluded. Citations provide annotation number. Numerals are filed as though they were spelled out. Initial articles in all languages are ignored in filing. Hyphenated words are considered as separate words in filing.

A

B

C

D

E

F

G

H

I

J

K

L

M

N

O

P

Q

R

S

T

U

V

W

Y

Z

Title Index

Entries are given for all titles, including separately titled essays within monographs. Citations provide annotation number. Numerals are filed as though they were spelled out. Initial articles are ignored in filing. Hyphenated words are considered as separate words in filing.

B

C

D

E

F

G

H

I

J

K

L

M

N

O

P

Q

R

T

U

V

W

Y